ENGINEERS' PRACTICAL DATABOOK

First Edition

ENGINEERS' PRACTICAL DATABOOK
First Edition in SI Units

This Data Book is provided for the teaching of engineering and conforms to typical teaching structure for selected modules within HNC, HND, Foundation Degree and Bachelor's Degree qualifications in Engineering.

Credit: Material properties derived or calculated from various sources, including *Materials Handbook* (ASM Vol.2, 1979), *Granta Design* (2018), *Thermodynamics: An Engineering Approach* (Çengel & Boles, 6th ed., 2007), *www.engineeringtoolbox.com* (2018), *www.matweb.com* (2018). Unit Circle Angles by Gustavb is licenced under CC BY-SA 3.0. Avogadro constant (2018 definition) and atomic elements from IUPAC (2018). The author accepts no liability for any injuries or damages caused that may result from the reader's acting upon or using the content contained in the publication. Always consult a professional.

Whilst every care has been taken to include accurate information, the author would appreciate any corrections to be sent to EngineersDatabook@gmail.com, quoting the page number.

All formulae are written in SI units (m, kg, s) unless otherwise explicitly stated.

ISBN: 978-198-061934-5

Brief Contents

Contents

SYMBOLS

	Name	Uses
α	alpha	Angles, angular acceleration, thermal expansion coefficient
β	beta	Angles, coefficients
γ	gamma	Heat capacity ratio, kinematic viscosity, shear strain
Γ	Gamma	Circulation (fluid dynamics)
δ	delta	Difference, damping
Δ	Delta	Difference, determinant (matrix)
ε	epsilon	Strain, permittivity (electromagnetism), electromotive force (EMF), random error (regression), emissivity (thermodynamics)
η	eta	Efficiency
θ	theta	Angle, temperature (thermodynamics)
κ	kappa	Curvature $(= 1/R)$
λ	lambda	Thermal conductivity, wavelength, eigenvalue, parameter in the Poisson Distribution (mean, variance)
μ	mu	Friction coefficient, dynamic viscosity
ν	nu	Kinematic viscosity, Poisson's ratio
ρ	rho	Mass density, resistivity (electrical), curvature (*alternative to r*)
σ	sigma	Normal stress, standard deviation
τ	tau	Shear stress, torque, time constant (electronics)
φ	phi	Angles, heat flow, potential energy, magnetic flux
ψ	psi	Helix angle (gears), stream function (fluid dynamics)
ω	omega	Angular velocity
Ω	Omega	Electrical resistance (ohm)
ζ	zeta	Damping ratio
A		Area (m^2), current (A)
D		Diameter (m)
E		Young's modulus (Pa)
G		Shear modulus (Pa)
I		Mass moment of inertia $(kg\,m^2)$, area moment of inertia (m^4)
J		Polar moment of inertia (m^4)
R		Radius (m)
Q		Volume flow rate (m^3/s), heat flowrate (kW)
V		Volume (m^3), velocity (m/s)

UNITS OF MEASUREMENT

PREFIXES

Symbol	Prefix	Multiplication factor	
P	peta	1 000 000 000 000 000	10^{15}
T	tera	1 000 000 000 000	10^{12}
G	giga	1 000 000 000	10^{9}
M	mega	1 000 000	10^{6}
k	kilo	1 000	10^{3}
d	deci	0.1	10^{-1}
c	centi	0.01	10^{-2}
m	milli	0.001	10^{-3}
μ	micro	0.000 001	10^{-6}
n	nano	0.000 000 001	10^{-9}
p	pico	0.000 000 000 001	10^{-12}

S.I. UNITS

Symbol	Unit	Quantity	Dimension
m	metre	Length	[L]
kg	kilogram	Mass	[M]
s	second	Time	[T]
A	ampere	Electric current	[A]
K	kelvin	Temperature	[θ]
cd	candela	Luminous intensity	[I]
mol	mole	Amount of substance	[N]

DERIVED UNITS

Quantity	Unit	Symbol	Base Units
Force	newton	N	$kg\, m\, s^{-2}$
Pressure and Stress	pascal	Pa	$kg\, m^{-1}\, s^{-2}$
Torque	newton-metre	N·m	$kg\, m^2\, s^{-2}$
Energy, Work, Heat	joule	J	$kg\, m^2\, s^{-2}$
Power	watt	W	$kg\, m^2\, s^{-3}$
Frequency	hertz	Hz	s^{-1}
Plane angle	radian	rad	$m\, m^{-1} = 1$
Solid angle	steradian	sr	$m\, m^{-2} = 1$
Luminous flux	lumen	lm	$sr\, cd$
Illuminance	lux	lx	$sr\, m^{-2}\, cd$
Kinematic Viscosity	stokes	St	$m^2\, s^{-1}$
Dynamic Viscosity	poiseuille	Pl	$kg\, m^{-1}\, s^{-1}$
Magnetic Flux	weber	Wb	$kg\, m^2\, s^{-2}\, A^{-1}$
Magnetic Flux Density	tesla	T	$kg\, s^{-2}\, A^{-1}$
Electrical Capacitance	farad	F	$s^4\, A^2\, m^{-2}\, kg^{-1}$
Electrical Charge	coulomb	C	$A \cdot s$
Electrical Conductance	siemens	S	$kg^{-1}\, m^{-2}\, s^3\, A^2$
Electrical Inductance	henry	H	$kg\, m^2\, s^{-2}\, A^{-2}$
Electrical Resistance	ohm	Ω	$kg\, m^2\, s^{-3}\, A^{-2}$
Potential difference/ Electromotive force	volt	V	$kg\, m^2\, A^{-1}\, s^{-3}$

MATHEMATICAL CONSTANTS

Symbol	Value	Description
e	$2.718\,281\,828\,45\ldots$	Base of the natural logarithm
i	$\sqrt{-1}$	Imaginary unit
π	$3.141\,592\,653\,59\ldots$	Ratio of circumference to diameter of a circle
$1\,rad$	$(180°/\pi) = 57.2857795\ldots°$	Radian

PHYSICAL CONSTANTS

Symbol	Value	Description
c	$299\,792\,458\ m\,s^{-1}$	Speed of light in vacuum
e	$1.602\,176\,565\ \times 10^{-19}\,C$	Elementary charge (e^+)
ε_0	$8.854\,187\,817 \times 10^{-12}\,m^{-3}\,kg^{-1}\,s^4\,A^2$	Electric constant (permittivity of free space, or vacuum permittivity)
μ_0	$4\pi \times 10^{-7}\,kg\,m\,s^{-2}\,A^{-2}$	Magnetic constant (permeability of free space, or vacuum permeability)
g	$9.806\,65\,m/s^2$	Standard gravity
G	$6.673\,84\ \times 10^{-11}\,N\,m^2\,kg^{-2}$	Gravitational constant
R_0	$8.314\,J\,mol^{-1}\,K^{-1}$	Universal gas constant
R_s	$287.058\,J\,kg^{-1}\,K^{-1}$	Specific gas constant for dry air
N_A	$6.022\,140\,76 \times 10^{23}\,mol^{-1}$	Avogadro constant

CONVERSION FACTORS

Dimension	Metric	Metric/Imperial
Length	$1\,m = 100\,cm$ $= 1\,000\,mm$	$1\,m = 3.2808\,ft$ $= 39.3701\,in$
Area	$1\,m^2 = 10^4\,cm^2$ $= 10^6\,mm^2$	$1\,m^2 = 1{,}550\,in^2$ $= 10.764\,ft^2$
Volume	$1\,m^3 = 1\,000\,L$ $= 10^6\,cm^3\,(cc)$	$1\,m^3 = 6.1024 \times 10^4\,in^3$ $= 35.315\,ft^3$ $= 219.97\,gal\,(UK)$ $1\,gallon\,(imperial)$ $= 4.546\,L$ $= 4.546 \times 10^{-3}\,m^3$
Volume flow rate	$1\,m^3/s = 60\,000\,L/min$ $= 10^6\,cm^3\,s^{-1}$	$1\,m^3/s = 15\,850\,gal/\text{min}$ $= 35.315\,ft^3/s$ $1m^3/s = 2\,118.9\,ft^3/min$ (cfm)
Mass	$1\,kg = 1\,000\,g$ $1\,000\,kg = 1\,tonne$	$1\,kg = 2.2046226\,lbm$ $1\,lbm = 0.45359237\,kg$ $1\,ounce = 28.3495\,g$ $1\,slug = 32.174\,lbm$ $= 14.5939\,kg$ $1\,short\,ton = 2{,}000\,lbm$ $= 907.1847\,kg$
Density	$1\,g/cm^3 = 1\,kg/L$ $= 1\,000\,kg/m^3$	$1\,g/cm^3 = 62.428\,lbm/ft^3$ $= 0.036127\,lbm/in^3$

Dimension	Metric	Metric/Imperial
Velocity	$1\ m/s = 3.60\ km/h$	$1m/s = 3.2808\ ft/s$ $= 2.237\ mi/h$ $1\ mi/h = 1.46667\ ft/s$ $= 1.6093\ km/h$ $1\ m/s = 1.94384\ knots$ $1\ mi/h = 0.868976\ knots$ $1\ km/h = 0.539957\ knots$
Acceleration	$1m/s^2 = 100\ cm/s^2$ $= 0.10197\ g$	$1\ m/s^2 = 3.2808\ ft/s^2$ $1\ ft/s^2 = 0.3048\ m/s^2$
Force	$1\ N = 1kg\ m/s^2$ $= 10^5\ dyn$ $1\ kgf = 9.80665\ N$	$1\ N = 0.22481\ lbf$ $1\ lbf = 32.174\ lbm \cdot ft/s^2$ $= 4.44822\ N$ $= 4.44822 \times 10^5\ dyn$
Pressure	$1\ Pa = 1\ N/m^2$ $1\ mm\ Hg = 0.1333\ kPa$ $1\ kPa = 10^3\ Pa$ $= 10^{-3}\ MPa$ $1\ bar = 100\ kPa$ $= 10^5\ Pa$ $1\ atm = 101.325\ kPa$ $= 1.01325\ bar$ $= 760\ mm\ Hg\ at\ 0°C$ $= 1.03323\ kgf/cm^2$ $1\ MPa = 1\ MN/m^2$ $= 1N/mm^2$	$1\ Pa = 1.45038 \times 10^{-4}\ psi$ $(i.e.\ lbf/in^2)$ $1\ psi = 144\ lbf/ft^2$ $= 6.894757\ kPa$ $1\ atm = 14.696\ psia$ $= 29.92\ in\ Hg\ @30°F$ $1\ in\ Hg = 3.387\ kPa$

Dimension	Metric	Metric/Imperial
Energy	$1\ kJ = 1\ 000\ J$ $= 1\ 000\ N \cdot m$ $= 1\ kPa \cdot m^3$ $1\ kWh = 3\ 600\ kJ$ $= 3.6 \times 10^6\ J$ $1\ cal = 4.1868\ J$ $1\ Cal = 4.1868\ kJ$	$1\ kJ = 0.94782\ Btu$ $1\ Btu = 1.055056\ kJ$ $= 5.40395\ psia \cdot ft^3$ $= 778.169\ lbf \cdot ft$ $1\ kWh = 3412.14\ Btu$ $1\ therm = 10^5\ Btu$ $= 1.055 \times 10^5\ kJ$
Specific energy	$1\ kJ/kg = 1\ 000\ m^2/s^2$	$1\ Btu/lbm = 25\ 037 ft^2/s^2$ $= 2.326\ kJ/kg$ $1\ kJ/kg = 0.430\ Btu/lbm$
Power	$1\ W = 1\ J/s$ $1\ kW = 1\ 000\ W$ $= 1.341\ hp$ $1\ hp = 745.7\ W$ (mechanical) $1\ hp = 746\ W$ (electrical)	
Temperature	$a\ K = (a - 273.15)\ ^\circ C$	$a\ ^\circ C = (1.8a + 32)\ ^\circ F$
Heat flux	$1\ W/cm^2 = 10^4\ W/m^2$	$1 W\ m^{-2} = 0.3171\ btu\ h^{-1} ft^{-2}$
Dynamic viscosity	$1\ P$ (poise) $= 0.1\ Pl$ (0.1 poisseuille) $= 0.1\ Pa \cdot s$ $= 0.1\ kg\ m^{-1}\ s^{-1}$ $1\ cP$ (centipoise) $= 1\ mPa \cdot s = 10^{-3} Pl$ $= 10^{-3}\ kg\ m^{-1}\ s^{-1}$	

1. MATHEMATICS

1.1. ALGEBRA

Laws of Exponents

$a^0 = 1 \qquad (a \neq 0)$

$a^1 = a$

$a^m a^n = a^{m+n}$

$(a^m)^n = a^{mn}$

$(ab)^n = a^n b^n$

$\frac{a^n}{a^m} = a^{n-m}$

$a^{-n} = \frac{1}{a^n} \quad (a \neq 0)$

$\left(\frac{a}{b}\right)^{-n} = \left(\frac{b}{a}\right)^n = \frac{b^n}{a^n}$

$a^n = \frac{1}{a^{-n}}$

$\sqrt[n]{a} = a^{\frac{1}{n}}$

$a^{n/m} = \sqrt[m]{a^n} = \left(\sqrt[m]{a}\right)^n$

e (the base of the natural logarithm)

$y = e^x$ *is the solution of* $\frac{dy}{dx} = y$

(with boundary conditions $y = 1$ at $x = 0$)

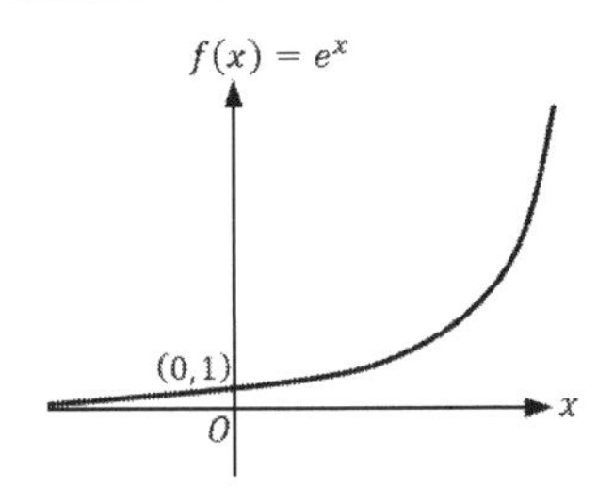

Decay (Decreasing):

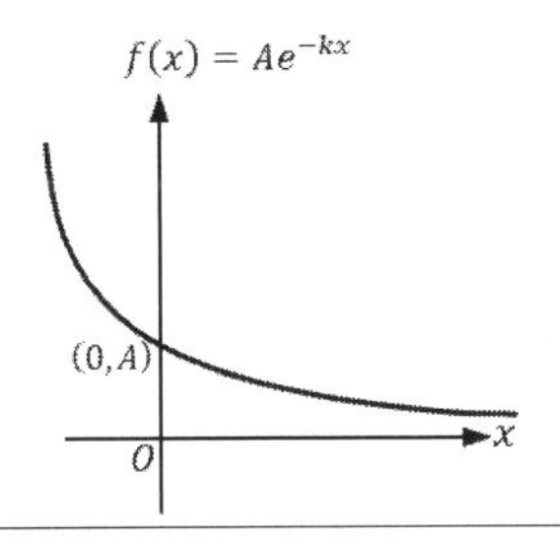

Decay (Increasing):

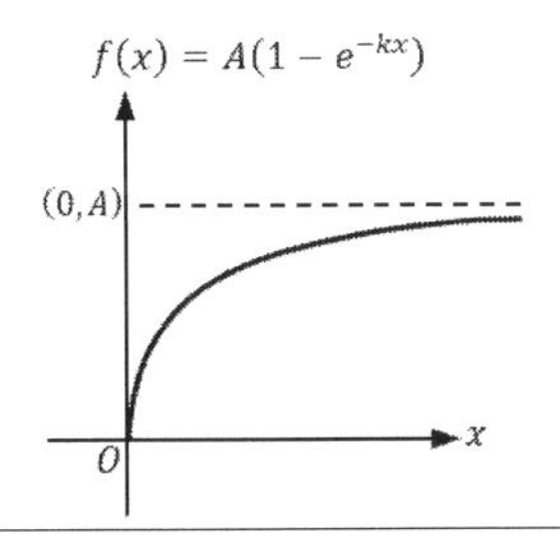

Laws of Logarithms

$$if\ N = a^x,\ then\ \ x = \log_a N$$

$$x = a^y \Leftrightarrow y = \log_a x$$

Exponents and logarithms are inverse functions.

$\log 1 = \log_{10} 1 = 0$

$\log_a (1) = 0$ LOGARITHM OF 1

$\log_a (a) = 1$ LOGARITHM OF THE BASE

The **common logarithm** is base 10.
E.g. $\log(0.01) = -2,\ \log(10) = 1,\ \log(100) = 2,\ \log(1000) = 3\ etc.$

$\log(x \cdot y) = \log x + \log y$ PRODUCT RULE

$\log \frac{x}{y} = \log x - \log y$ QUOTIENT RULE

$\log x^n = n \log x$ POWER RULE

$\log_a (b) = \frac{1}{\log_b (a)}$ BASE SWITCH

$\log_a (x) = \frac{\log_b (x)}{\log_b (a)}$ CHANGE OF BASE

The **natural logarithm** is base e, where $e \approx 2.71828$.

$\ln y = \log_e y$ NATURAL LOGARITHM

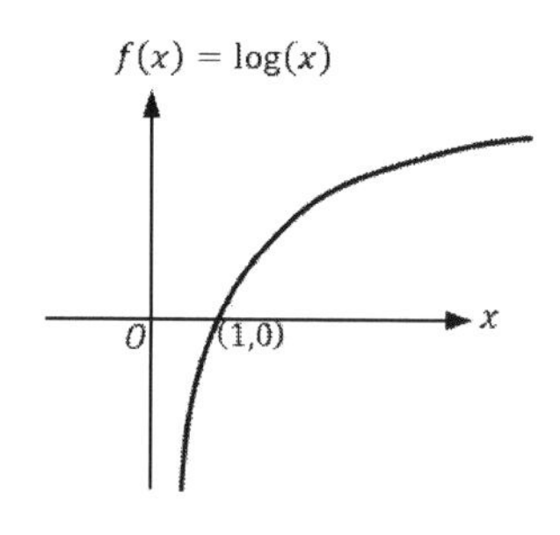

Polynomials

$$p(x) = a_n x^n + a_{n-1} x^{n-1} + \dots + a_2 x^2 + a_1 x^1 + a_0$$

A polynomial function $p(x)$ is the linear sum of terms containing the positive integer powers of x, each with a constant coefficient (e.g. $2x^3 + 3x^2 - 4x + 0.5 = 9$).
The **degree** of a polynomial is its highest power of x.
$p(x)$ above is a degree n polynomial. A degree 1 polynomial is **linear** (e.g. $6.5x + 0.75$).
A degree 2 polynomial is a **quadratic** (e.g. $2x^2 - 6x + 10$). A degree 3 polynomial is **cubic**. A degree 4 polynomial is **quartic**.

Factor Theorem

If $p(\alpha) = 0$ then $x - \alpha$ is a factor of the polynomial $p(x)$.

Example: *For* $p(x) = x^3 + x^2 + x + 1$
$p(-1) = (-1)^3 + (-1)^2 + (-1) + 1 = -1 + 1 - 1 + 1 = 0$ therefore (x+1) is a factor of $p(x)$.

Quadratic Equation

$ax^2 + bx + c = 0,\ \ a \neq 0$ *has solutions* x_1, x_2, *where*

$$x_{1,2} = \frac{-b \pm \sqrt{b^2 - 4ac}}{2a}$$

$$x_1 + x_2 = \frac{-b}{a}, \quad x_1 x_2 = \frac{c}{a}$$

The discriminant of the quadratic is $b^2 - 4ac$.
If the discriminant is positive, there exists two real roots, x_1 and x_2.
If the discriminant is zero, there exists two repeated roots i.e. $x_1 = x_2$.
If the discriminant is negative, there exists two complex roots $\alpha + j\omega$ and $\alpha - j\omega$.

$f(x) = ax^2 + bx + c$

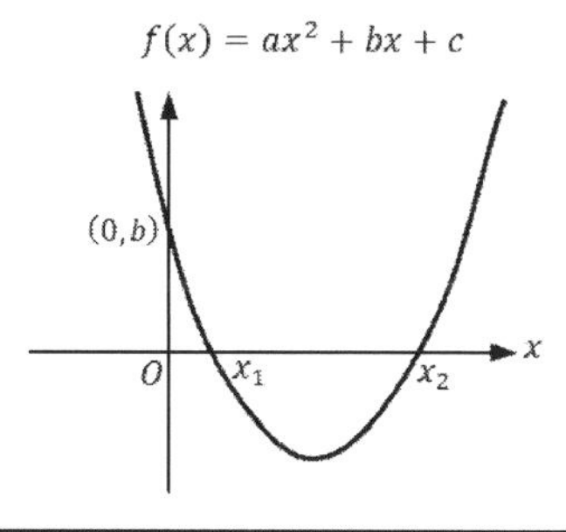

If $f(a) = 0$ then a is a root of the quadratic equation, and $(x - a)$ is a factor, i.e. $f(x)$ is divisible by $(x - a)$ with no remainder. Roots of the polynomial equation correspond to points where the curve crosses the x axis.

Partial Fractions

Expression	Form of partial fractions
Linear factors	
$\dfrac{f(x)}{(x+a)(x+b)(x+c)}$	$\dfrac{A}{(x+a)}+\dfrac{B}{(x+b)}+\dfrac{C}{(x+c)}$
Repeated linear factors	
$\dfrac{f(x)}{(x+a)^3}$	$\dfrac{A}{(x+a)}+\dfrac{B}{(x+a)^2}+\dfrac{C}{(x+a)^3}$
Quadratic factors	
$\dfrac{f(x)}{(ax^2+bx+c)(x-d)}$	$\dfrac{Ax+B}{(ax^2+bx+c)}+\dfrac{C}{(x-d)}$
General example	
$\dfrac{f(x)}{(x^2+a)(x+b)^2(x+c)}$	$\dfrac{Ax+B}{(x^2+a)}+\dfrac{C}{(x+b)}+\dfrac{D}{(x+b)^2}+\dfrac{E}{(x+c)}$

Partial fraction decomposition is useful in computing indefinite integrals where there is a polynomial in the denominator. It is also used in Laplace transforms, and in finding the solution to some differential equations.

1.2. SEQUENCES AND SERIES

Sum of First n Natural Numbers

The sum of the first n natural numbers $(1 + 2 + 3 + \ldots + n)$ is

$$S_n = \sum_{r=1}^{n} r = \frac{n}{2}(n+1)$$

Sum of First n Squared Natural Numbers

The sum of the first n^2 natural numbers $(1^2 + 2^2 + 3^2 + \ldots + n^2)$ is

$$S_{n^2} = \sum_{r=1}^{n} r^2 = \frac{n(n+1)(2n+1)}{6}$$

Sum of Arithmetic Progression

A sequence is a list of objects, numbers, or variables.

The general form of an arithmetic sequence is

$$a, (a+d), (a+2d), (a+3d) \ldots (a+(n-2)d) + (a+(n-1)d)$$

Summation is the addition of a sequence of numbers

The sum of n arithmetic terms is given by

$$S_n = \frac{n}{2}(a+l)$$

The last number in an arithmetic sequence is $l = a + (n-1) \cdot d$

$$S_n = \frac{n}{2}(2a + (n-1)d)$$

$a = first\ number\ in\ sequence$
$l = last\ number\ in\ sequence$
$d = difference\ between\ consecutive\ numbers\ in\ the\ arithmetic\ sequence$
$n = total\ number\ of\ objects\ in\ sequence$

Sum of Geometric Progression

The general form of a geometric sequence is

$$a, ar, ar^2, ar^3, \ldots, ar^{n-1}, ar^n$$

The nth term is

$$a_n = ar^{n-1}$$

The sum of n geometric terms is given by

$$S_n = \frac{a(1 - r^n)}{1 - r}, \ r \neq 1$$

The sum of a converging geometric series when $-1 < r < 1$:

$$S_\infty = \sum_{n=1}^{\infty} ar^{n-1} = \frac{a}{1 - r}, \ -1 < r < 1$$

$a = first\ term\ (i.e.\ the\ scale\ factor)$
$r = common\ ratio$

1.3. COORDINATE SYSTEMS

$P = (x_i, y_i)$ CARTESIAN

$P = r\angle\theta$ POLAR

$P = (r, \theta, z)$; *radius, azimuth, height* CYLINDRICAL

$P = (r, \theta, \varphi)$; *radius, azimuth, inclination* SPHERICAL

> A point in R^2 space may be represented as (x, y) or $r\angle\theta$.
> A point in R^3 space may be represented (x, y, z), (r, θ, z) or (r, θ, φ).

POLAR TO CARTESIAN

$x = r\cos\theta,\ y = r\sin\theta$

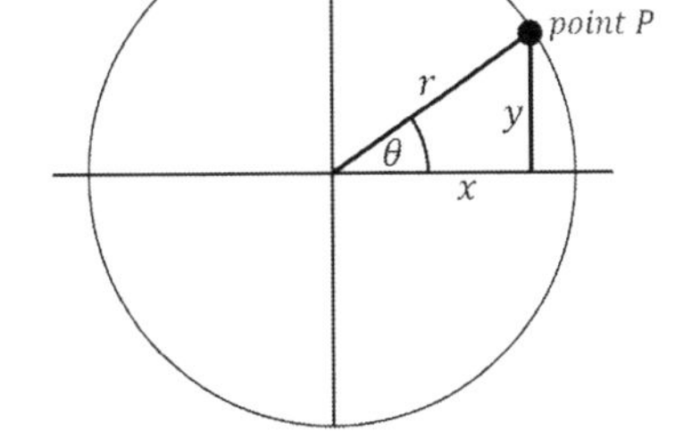

CARTESIAN TO POLAR

$$r = \sqrt{x^2 + y^2}$$

$$\theta = \arccos\left(\frac{x}{r}\right) = \arcsin\left(\frac{y}{r}\right) = \arctan\left(\frac{y}{x}\right)$$

CARTESIAN TO CYLINDRICAL

$$r = \sqrt{x^2 + y^2}$$

$$\theta = \arctan\left(\frac{y}{x}\right)$$

$$z = z$$

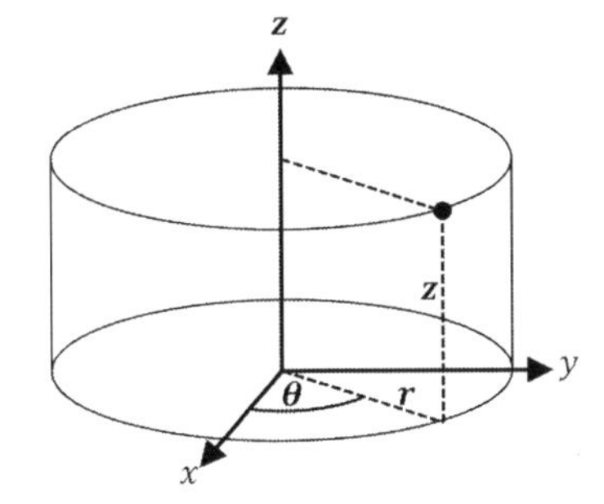

CARTESIAN TO SPHERICAL

$$r = \sqrt{x^2 + y^2 + z_2}$$

$$\theta = \arctan\left(\frac{y}{x}\right)$$

$$\varphi = \arccos\left(\frac{z}{r}\right)$$

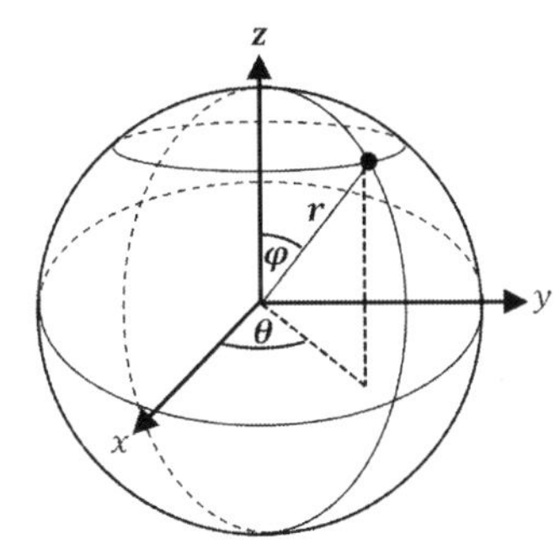

1.4. COMPLEX NUMBERS

$i^2 = j^2 = -1$ (note: Mathematicians prefer *i*, Engineers *j*)

$i = \sqrt{-1},\ i\sqrt{a} = \sqrt{-a}$

Cartesian Form

$z = x + iy$

$\bar{z} = z^* = x - iy$ COMPLEX CONJUGATE

A complex conjugate results from reflecting a point in the complex plane about the real axis. To find the conjugate of a complex number all you have to do is change the sign between the two terms in the denominator.

Properties of Complex Numbers

$z_1 + z_2 = (x_1 + x_2) + i(y_1 + y_2)$

$z_1 - z_2 = (x_1 - x_2) + i(y_1 - y_2)$

To add or subtract complex numbers, evaluate real and imaginary parts separately.

$z_1 \cdot z_2 = (x_1 x_2 - y_1 y_2) + i(x_1 y_2 + x_2 y_1)$

Multiply by expanding $(x_1 + iy_1)(x_2 + iy_2)$, then using the definition $i^2 = -1$.

$$\frac{1}{x + iy} = \frac{(x - iy)}{(x + iy)(x - iy)} = \frac{x - iy}{x^2 + y^2}$$

To divide by a complex number, first multiply top and bottom by the conjugate.

Polar Form

$z = re^{i\theta} = r(\cos\theta + i\sin\theta) = r\ cis\theta$

$\bar{z} = z^* = re^{-i\theta} = r(\cos\theta - i\sin\theta)$ COMPLEX CONJUGATE

A point in C space may be represented as $x + iy$ *or* $re^{i\theta}$ *or* $r\angle\theta$

$z^n = r^n \angle n\theta$

$z_1 \cdot z_2 = r_1 e^{i\theta_1} \cdot r_2 e^{i\theta_2} = r_1 \cdot r_2 e^{i(\theta_1 + \theta_2)}$

To multiply complex numbers, multiply the moduli (lengths) and add the arguments (angles).

Modulus

$mod(z) = |z| = \sqrt{x^2 + y^2}$

$|z \cdot \bar{z}| = r^2$

Argument

$\arg(z) = \theta = \arctan\left(\frac{y}{x}\right)$

De Moivre's Theorem

$(\cos\theta + i\sin\theta)^n = (\cos n\theta + i\sin n\theta)$

$(r\,e^{i\theta})^n = r^n e^{in\theta} = r^n(\cos\theta + i\sin\theta)^n = r^n(\cos n\theta + i\sin n\theta)$

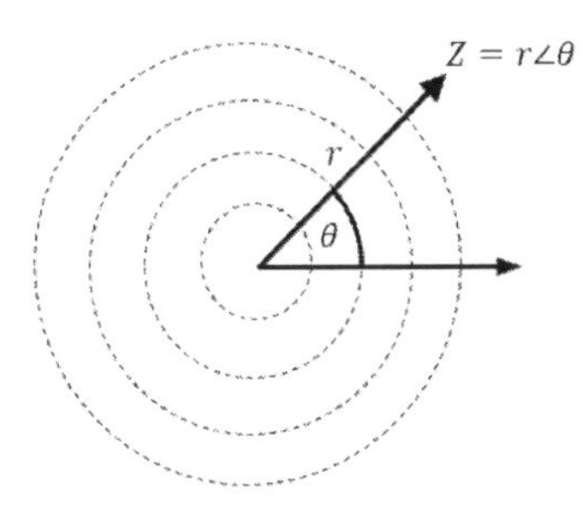

Polar form

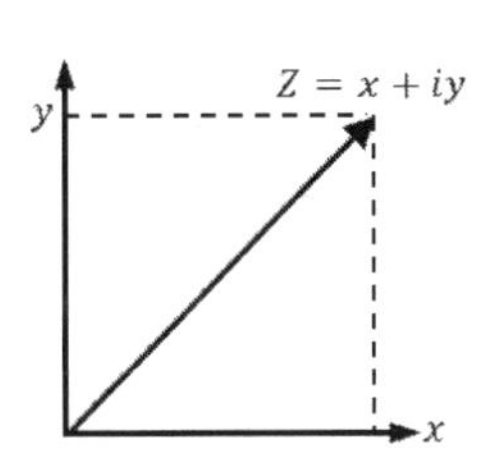

Cartesian form

1.5. POWER SERIES

$$e^x = \sum_{k=0}^{\infty} \frac{x^k}{k!} = 1 + \frac{x}{1!} + \frac{x^2}{2!} + \frac{x^3}{3!} + \ldots, \quad -\infty < x < \infty$$

$$\sin x = \sum_{k=0}^{\infty} (-1)^k \frac{x^{2k+1}}{(2k+1)!} = x - \frac{x^3}{3!} + \frac{x^5}{5!} + \frac{x^7}{7!} + \ldots$$

$$\cos x = \sum_{k=0}^{\infty} (-1)^k \frac{x^{2k}}{(2k)!} = 1 - \frac{x^2}{2!} + \frac{x^4}{4!} + \frac{x^6}{6!} + \ldots$$

$$\tan x = x + \frac{x^3}{3} + \frac{2x^5}{15} + \frac{17x^7}{315} + \frac{62x^9}{2835} + \ldots$$

$$\arctan x = x - \frac{x^3}{3} + \frac{x^5}{5} - \frac{x^7}{7} + \ldots, \quad -1 \le x \le 1$$

$$(1+x)^n = 1 + \frac{nx}{1!} + \frac{n(n-1)x^2}{2!} + \frac{n(n-1)(n-2)x^3}{3!} + \ldots + x^n$$

$$\frac{1}{1+x} = 1 - x + x^2 - x^3 + x^4 + \ldots + (-1)^n x^n + \ldots, \quad -1 < x < 1$$

$$\frac{1}{1-x} = 1 + x + x^2 + x^3 + \ldots, \quad -1 < x < 1$$

Taylor Series

$$f(x) = f(a) + f^{'}(a)(x - a) + \frac{f^{''}(a)}{2!}(x - a)^2 + \frac{f^3(a)}{3!}(x - a)^3 + \dots$$

$$+ \frac{f^n(a)}{n!}(x - a)^n + \dots$$

> If $f(x)$ is a continuously differentiable function about the point $x = a$, then the Taylor Series expansion gives $f(x)$ in terms of a power series.

Maclaurin Series

> A Maclaurin Series is a Taylor series expansion of an 'infinitely differentiable' function $f(x)$ about the point $x = 0$.

$$f(x) = f(0) + f^{'}(0)x + \frac{f^{''}(0)}{2!}x^2 + \frac{f^3(0)}{3!}x^3 + \dots + \frac{f^n(0)}{n!}x^n + \dots$$

> The first few terms may be used as a good approximation for values near to $x = 0$.

1.6. FOURIER SERIES

For a function $f(x)$ periodic on the interval $[-L, L]$:

$$f(x) = \frac{1}{2}a_0 + \sum_{n=1}^{\infty}\left(a_n\cos\frac{n\pi x}{L} + b_n\sin\frac{n\pi x}{L}\right)$$

$$a_0 = \frac{1}{\pi}\int_{-L}^{L} f(x)dx$$

$$a_n = \frac{1}{L}\int_{-L}^{L} f(x)\cos\left(\frac{n\pi x}{L}\right)dx$$

$$b_n = \frac{1}{L}\int_{-L}^{L} f(x)\sin\left(\frac{n\pi x}{L}\right)dx$$

Fourier Series

Rectangular Wave (cosines)

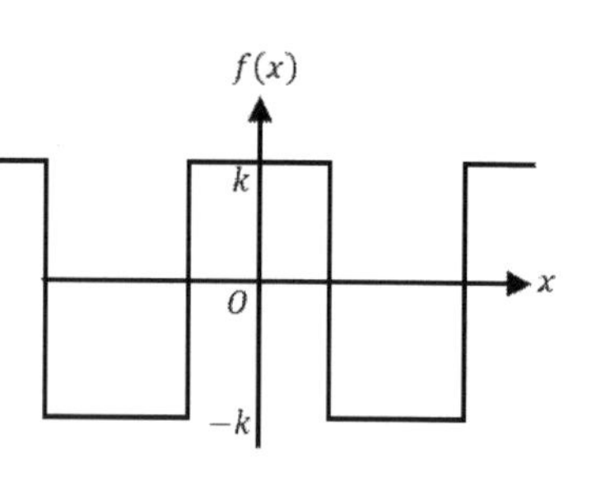

$$f(x) = \begin{cases} -k & L/2 < |x| < 3L/2 \\ +k & |x| < L/2 \end{cases}$$

$$f(x) = \frac{4k}{\pi}\left[\cos\frac{\pi x}{L} - \frac{1}{3}\cos\frac{3\pi x}{L} + \frac{1}{5}\cos\frac{5\pi x}{L} - \ldots\right]$$

Rectangular Wave (sines)

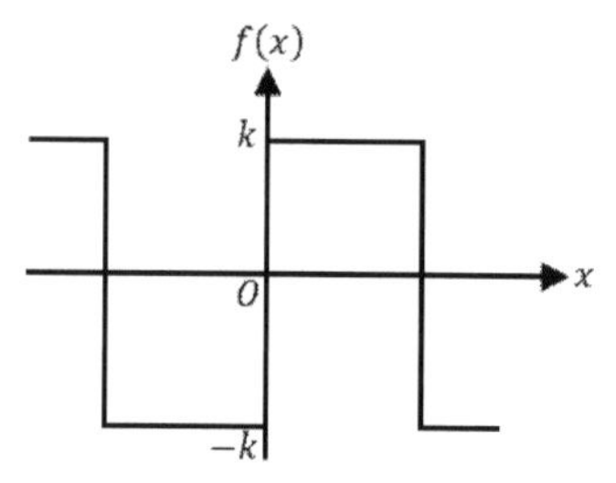

$$f(x) = \begin{cases} -k & -L < |x| < 0 \\ +k & 0 < |x| < L \end{cases}$$

$$f(x) = \frac{4k}{\pi}\left[\sin\frac{\pi x}{L} + \frac{1}{3}\sin\frac{3\pi x}{L} + \frac{1}{5}\sin\frac{5\pi x}{L} + \ldots\right]$$

Triangular Wave

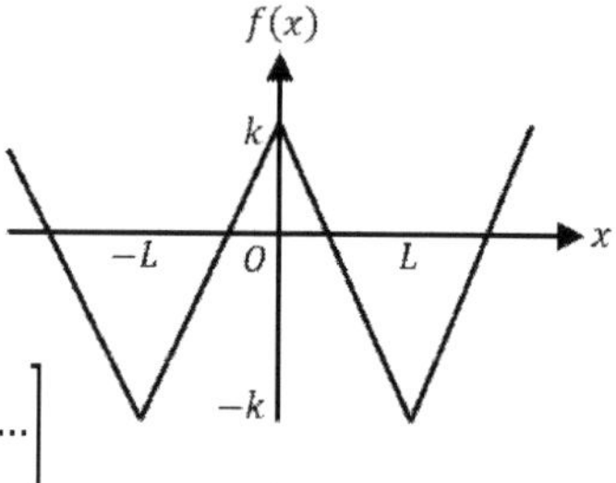

$$f(x) = \begin{cases} k(1 + 2x/L) & -L < |x| < 0 \\ k(1 - 2x/L) & 0 < |x| < L \end{cases}$$

$$f(x) = \frac{8k}{\pi}\left[\cos\frac{\pi x}{L} + \frac{1}{3^2}\cos\frac{3\pi x}{L} + \frac{1}{5^2}\cos\frac{5\pi x}{L} + \ldots\right]$$

Saw Tooth Wave

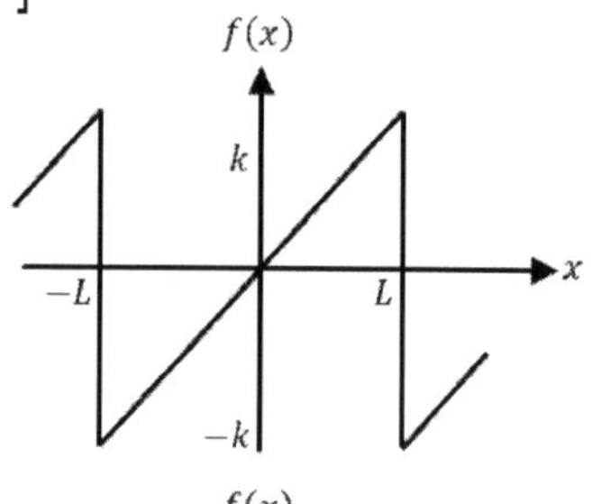

$$f(x) = kx/L$$

$$f(x) = \frac{2k}{\pi}\left[\sin\frac{\pi x}{L} - \frac{1}{2}\sin\frac{2\pi x}{L} - \frac{1}{3}\sin\frac{3\pi x}{L} + \ldots\right]$$

Full Wave Rectification

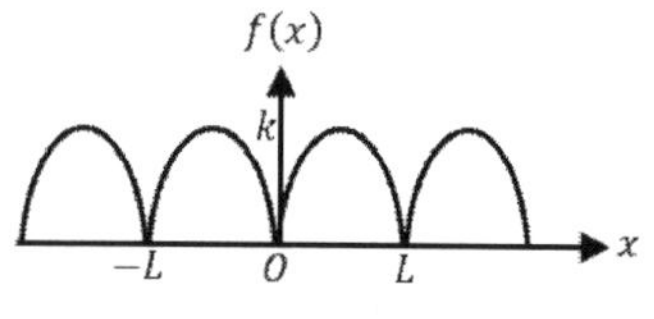

$$f(x) = k\left|\sin\left(\frac{\pi x}{L}\right)\right|$$

$$f(x) = \frac{4k}{\pi}\left[\frac{1}{2} - \frac{1}{1\times 3}\cos\left(\frac{2\pi x}{L}\right) - \frac{1}{3\times 5}\cos\left(\frac{4\pi x}{L}\right) - \frac{1}{5\times 7}\cos\left(\frac{6\pi x}{L}\right) - \ldots\right]$$

1.7. TRIGONOMETRY

Definitions

$$\sin\theta = \frac{opp}{hyp} \quad and \quad \theta = \arcsin\left(\frac{opp}{hyp}\right)$$

$$\cos\theta = \frac{adj}{hyp} \quad and \quad \theta = \arccos\left(\frac{adj}{hyp}\right)$$

$$\tan\theta = \frac{opp}{adj} \quad and \quad \theta = \arctan\left(\frac{opp}{adj}\right)$$

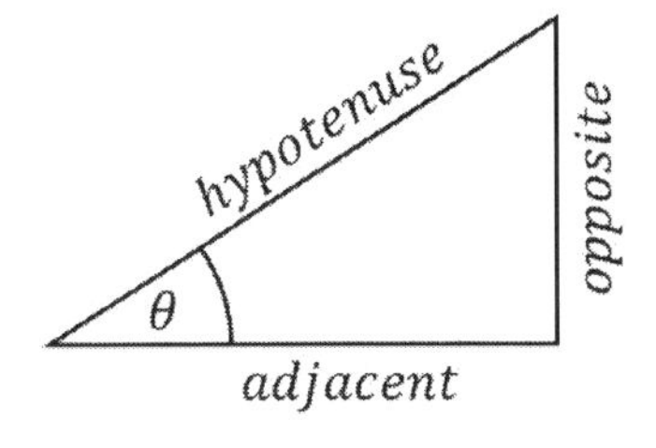

$R\cos\theta$ is a projection of a hypotenuse of length R onto the x axis;
$R\sin\theta$ is a projection of a hypotenuse of length R onto the y axis.

$$\csc\theta = \frac{1}{\sin\theta}, \ \sec\theta = \frac{1}{\cos\theta}, \ \cot\theta = \frac{1}{\tan\theta}$$

Unit Circle Identities

$\cos^2\theta + \sin^2\theta = 1$	$1 + \tan^2\theta = \sec^2\theta$	$1 + \cot^2\theta = cosec^2\,\theta$

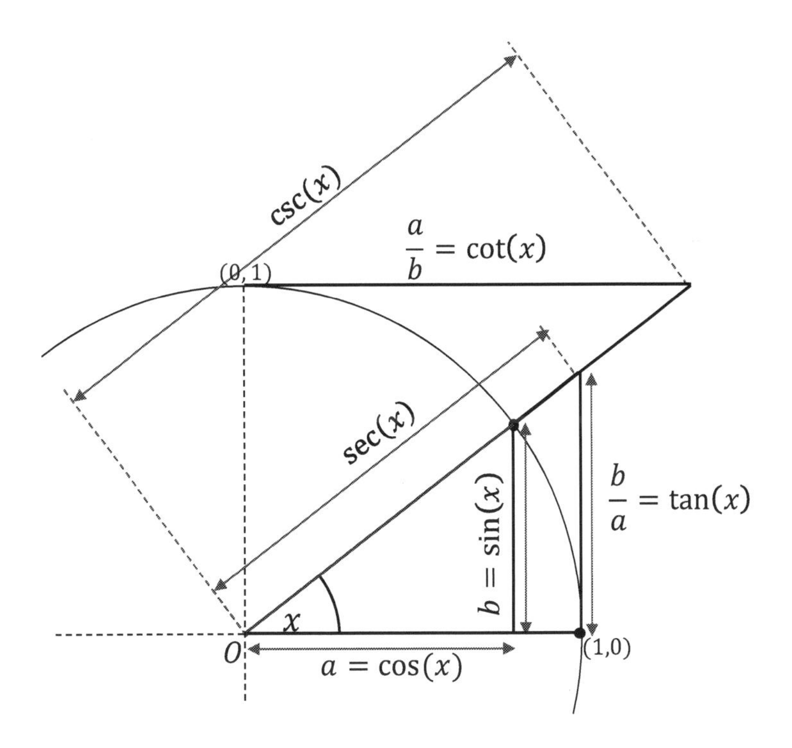

Unit Circle Angles

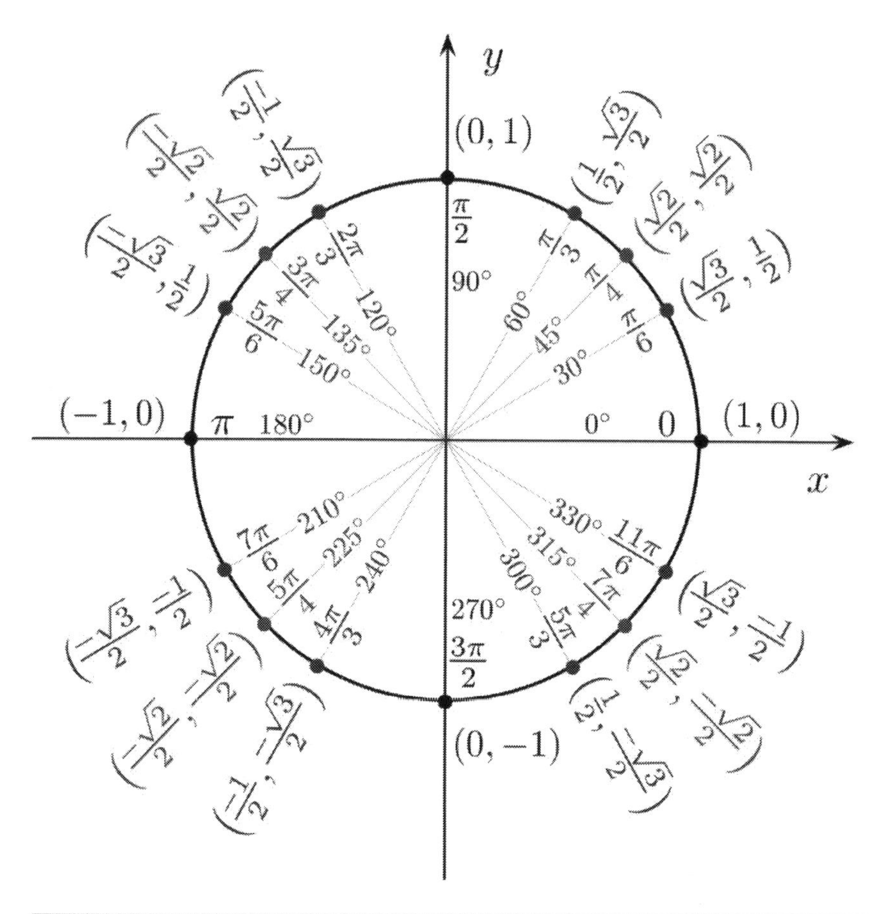

x (*rad*)	**0**	$\frac{\pi}{6}$	$\frac{\pi}{4}$	$\frac{\pi}{3}$	$\frac{\pi}{2}$
x (*deg*)	0°	30°	45°	60°	90°
sin x	0	$\frac{1}{2}$	$\frac{1}{\sqrt{2}} = \frac{\sqrt{2}}{2}$	$\frac{\sqrt{3}}{2}$	1
cos x	1	$\frac{\sqrt{3}}{2}$	$\frac{1}{\sqrt{2}} = \frac{\sqrt{2}}{2}$	$\frac{1}{2}$	0
tan x	0	$\frac{1}{\sqrt{3}}$	1	$\sqrt{3}$	∞

Trigonometric Identities

Sums and Differences Formula

$$\sin(A) \pm \sin(B) = 2\sin\left(\frac{1}{2}(A \pm B)\right)\cos\left(\frac{1}{2}(A \mp B)\right)$$

$$\sin(A \pm B) = \sin(A)\cos(B) \pm \cos(A)\sin(B)$$

$$\cos(A \pm B) = \cos(A)\cos(B) \mp \sin(A)\sin(B)$$

$$\tan(A \pm B) = \frac{\tan(A) \pm \tan(B)}{1 \mp \tan(A)\tan(B)}$$

Double Angle Formula

$$\cos(2A) = \cos^2(A) - \sin^2(A) = 2\cos^2(A) - 1 = 1 - 2\sin^2(A)$$

$$\sin(2A) = 2\sin(A)\cos(A)$$

$$\tan(2A) = \frac{2\tan(A)}{1 - \tan^2(A)}$$

$$\cos^2(A) = \frac{1}{2}(1 + \cos(2A))$$

$$\sin^2(A) = \frac{1}{2}(1 - \cos(2A))$$

Half-Angle Formula

$$\cos\left(\frac{B}{2}\right) = \pm\sqrt{\frac{1 + \cos(B)}{2}}$$

$$\sin\left(\frac{B}{2}\right) = \pm\sqrt{\frac{1 - \cos(B)}{2}}$$

Sinusoidal Waveforms

The general form of a sinusoidal wave may be given by the expression:

$$y = R\sin(\omega t + \theta)$$

$|R| = amplitude\ (the\ maximum\ value\ of\ y)$
$\omega = angular\ frequency\ [rad\ s^{-1}]$
$t = time\ [s]$
$\theta = phase\ angle\ [rad]$

$$phase\ shift = \frac{-\theta}{\omega}$$ PHASE SHIFT

$phase\ shift = horizontal\ shift\ [s]$

$$T = \frac{1}{f} = \frac{2\pi}{\omega}$$ PERIOD

$T = period\ [s]$

$$f = \frac{1}{T} = \frac{\omega}{2\pi}$$ FREQUENCY

$f = frequency\ [Hz]$
$T = periodic\ time\ [s]$

$$\lambda = \frac{v}{f} = vT = \frac{2\pi v}{\omega}$$ WAVELENGTH

$\lambda = wavelength\ [m]$
$v = phase\ speed\ [m\ s^{-1}]$
$f = frequency\ [Hz]$

The most common periodic signal waveforms that are used in Electrical and Electronic Engineering for example are the Sinusoidal Waveforms (based on a sine or cosine function). However, an alternating AC waveform can also take other forms, such as Complex Waves, Square Waves or Triangular Waves.

Sine and Cosine Wave Relationships

$$\cos(\omega t + \theta) = \sin\left(\omega t + \theta + \frac{\pi}{2}\right)$$

$$\sin(\omega t + \theta) = \cos\left(\omega t + \theta - \frac{\pi}{2}\right)$$

Expansion of *R sin(ωt+θ)*

$$R\sin(\omega t+\theta) = a\sin(\omega t) + b\cos(\omega t)$$

where:

$a = R\cos\theta$

$b = R\sin\theta$

$R = \sqrt{a^2 + b^2}$

$\theta = \arctan\left(\frac{b}{a}\right)$

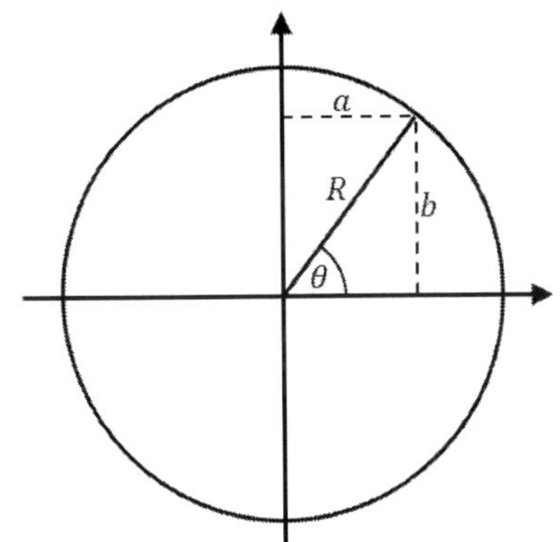

Use Sums and Differences formulae to expand $R\sin(\omega t+\theta)$.
Compare coefficients to find a, b.

Area of a Triangle

$$Area = \frac{1}{2}a \cdot b \cdot \sin(C) = \frac{1}{2}a \cdot c \cdot \sin(B) = \frac{1}{2}b \cdot c \cdot \sin(A)$$

Law of Sines and Cosines

$$\frac{a}{\sin(A)} = \frac{b}{\sin(B)} = \frac{c}{\sin(C)}$$

$c^2 = a^2 + b^2 - 2ab\cos(A)$ UNKNOWN LENGTH

$A = \operatorname{acos}\left(\frac{b^2 + c^2 - a^2}{2bc}\right)$ UNKNOWN ANGLE

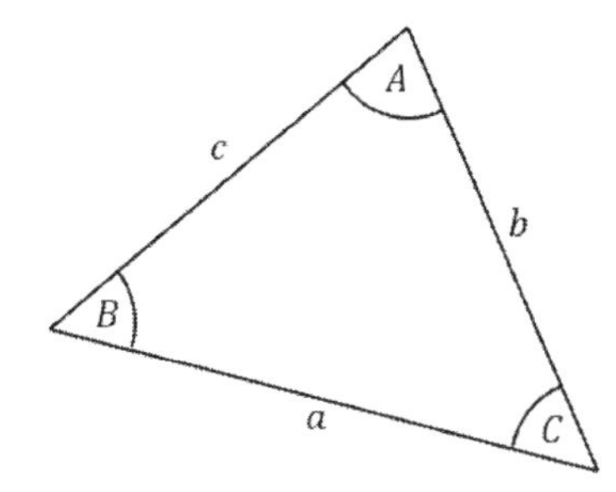

1.8. HYPERBOLIC FUNCTIONS

$$\sinh x = \frac{e^x - e^{-x}}{2}$$

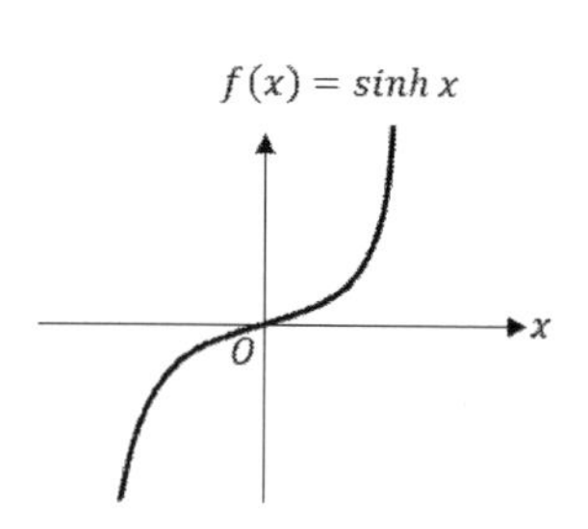

$$\cosh x = \frac{e^x + e^{-x}}{2}$$

$$\tanh x = \frac{e^x - e^{-x}}{e^x + e^{-x}} = \frac{e^{2x} - 1}{e^{2x} + 1}$$

A chain suspended from equal heights at its ends and dropping at $x = 0$ to its lowest height $y = a$ is given by the equation $y = a\cosh(x/a)$.

$$\operatorname{csch} x = \frac{1}{\sinh x} = \frac{2}{e^x - e^{-x}}$$

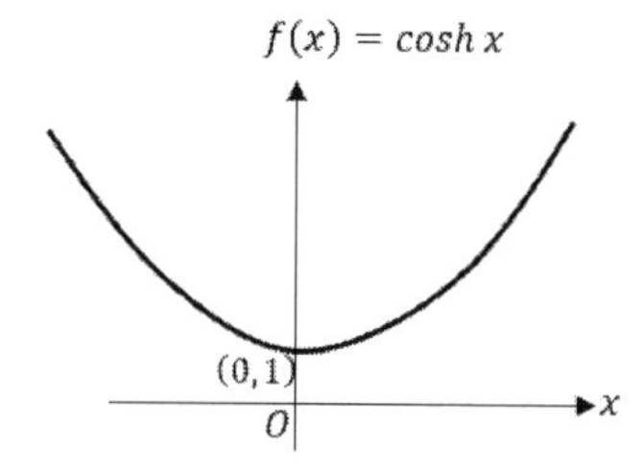

$$\operatorname{sech} x = \frac{1}{\cosh x} = \frac{2}{e^x + e^{-x}}$$

$$\coth x = \frac{1}{\tanh x} = \frac{e^x + e^{-x}}{e^x - e^{-x}} = \frac{e^{2x} + 1}{e^{2x} - 1}$$

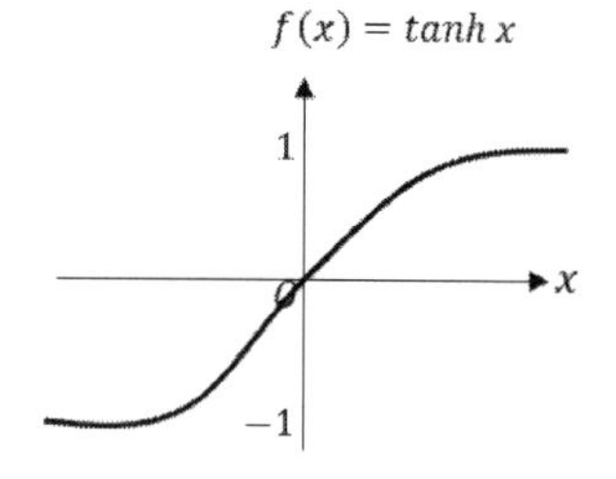

$$\cosh x + \sinh x = e^x$$

$$\cosh x - \sinh x = e^{-x}$$

$$\cosh^2 x - \sinh^2 x = 1$$

$$\operatorname{sech}^2 x = 1 - \tanh^2 x$$

1.9. VECTORS

Notation

$\begin{pmatrix} a_x & a_y & a_z \end{pmatrix}$

1×3 ROW VECTOR

$\begin{pmatrix} b_x \\ b_y \\ b_z \end{pmatrix}$

3×1 COLUMN VECTOR

$|\vec{a}| = \sqrt{a_x^2 + a_y^2 + a_z^2}$

MAGNITUDE

$\vec{a}/a = \hat{a}$

UNIT VECTOR

$\vec{a} = |\vec{a}|\hat{a} = a\hat{a}$

MAGNITUDE DIRECTION

$\hat{\imath},\ \hat{\jmath},\ \hat{k}$

UNIT VECTORS IN x, y, z

> A unit vector is typically represented with a *'circumflex'* or hat, e.g. $\hat{\imath}$, $\hat{\jmath}$, $\hat{k}$, $\hat{e}_\theta$, $\hat{e}_r$, $\hat{e}_z$.

$\vec{a} = a_x\hat{\imath} + a_y\hat{\jmath} + a_z\hat{k}$

COMPONENTS (3D)

> Vectors may be distinguished from scalars using bold lowercase font, or accented with a bar, underline or arrow, e.g. ***a***, $\bar{v}$, $\underline{r}$, $\vec{\omega}$.

Vector Multiplication

$$\vec{a} \cdot \vec{b} = |\vec{a}||\vec{b}|\cos\theta$$
$$= a_x b_x + a_y b_y + a_z b_z$$

SCALAR PRODUCT

> The projection of vector $\vec{a}$ onto an axis is given by the scalar product of $\vec{a}$ and the unit vector along that axis, e.g. the components of $\vec{a}$ in x, y and z are $\vec{a}\cdot\hat{\imath}$, $\vec{a}\cdot\hat{\jmath}$, and $\vec{a}\cdot\hat{k}$.

$$\theta = \cos^{-1}\left(\frac{\vec{a}\cdot\vec{b}}{|\vec{a}||\vec{b}|}\right)$$

ANGLE BETWEEN VECTORS

> If a dot product of two vectors is zero, those vectors are perpendicular to each other.

$$\vec{a} \cdot \vec{a} = |\vec{a}|^2 = a_x^2 + a_y^2 + a_z^2$$

SCALAR PRODUCT IDENTITY

> Arrow, bar, or underline are interchangeable, and typically used in handwritten calculations to represent a vector e.g. $\vec{a}$, $\bar{v}$, $\underline{r}$.

$$\vec{a} \times \vec{b} = |\vec{a}||\vec{b}|\,\hat{n}\sin\theta$$

VECTOR PRODUCT

> Vector product conventionally follows the *right-hand rule,* i.e. using your right hand, if $\vec{a}$ were in the direction of your index finger, and $\vec{b}$ were in the direction of your middle finger naturally curled, then the vector $\vec{c} = \vec{a} \times \vec{b}$ would be at 90° to both $\vec{a}$ and $\vec{b}$, pointing in the direction of your thumb.

$$\theta = \operatorname{asin}\left(\frac{|\vec{a}\times\vec{b}|}{|\vec{a}||\vec{b}|}\right)$$

ANGLE BETWEEN VECTORS

$$\vec{a} \times \vec{b} = \begin{vmatrix} \hat{\imath} & \hat{\jmath} & \hat{k} \\ a_x & a_y & a_z \\ b_x & b_y & b_z \end{vmatrix}$$

VECTOR PRODUCT

$$= (A_y B_z - A_z B_y)\hat{\imath} + (A_z B_x - A_x B_z)\hat{\jmath} + (A_x B_y - A_y B_x)\hat{k}$$

$$\vec{a} \times \vec{b} = -\vec{b} \times \vec{a}$$

ANTICOMMUTATIVE LAW

$$\vec{a} \times (\vec{b} + \vec{c}) = (\vec{a} \times \vec{b}) + (\vec{a} \times \vec{c})$$

DISTRIBUTIVE LAW

> If two vectors $\vec{a}$ and $\vec{b}$ describe two adjacent edges of a parallelogram, the vector product gives the (signed) area of the shape.

Triple Products

$$\vec{a} \cdot (\vec{b} \times \vec{c}) = \vec{b} \cdot (\vec{c} \times \vec{a})$$

SCALAR TRIPLE PRODUCT

If three vectors $\vec{a}, \vec{b}, \vec{c}$ each describe the three edges of a parallelepiped, the scalar triple product gives the (signed) volume.

$$\vec{a} \times (\vec{b} \times \vec{c}) = \vec{b}(\vec{a} \cdot \vec{c}) - \vec{c}(\vec{a} \cdot \vec{b})$$

$$(\vec{a} \times \vec{b}) \times \vec{c} = \vec{b}(\vec{a} \cdot \vec{c}) - (\vec{c} \cdot \vec{b})\vec{a}$$

VECTOR TRIPLE PRODUCT; LAGRANGE'S FORMULA

Vector Calculus

$$\nabla = \hat{\imath}\frac{\partial}{\partial x} + \hat{\jmath}\frac{\partial}{\partial y} + \hat{k}\frac{\partial}{\partial z}$$

DEL OPERATOR

Scalar Field $\varphi(x, y, z)$

A scalar field associates a scalar value to every point in a space. The scalar may either be a (dimensionless) number or a physical quantity. Examples of scalar fields include pressure, temperature, humidity, gravitational potential, electric potential.

$$\nabla\varphi = grad\ \varphi = \hat{\imath}\frac{\partial\varphi}{\partial x} + \hat{\jmath}\frac{\partial\varphi}{\partial y} + \hat{k}\frac{\partial\varphi}{\partial z}$$

GRADIENT

Gradient of a scalar field φ gives the direction and magnitude of steepest ascent.

Vector Field $\vec{V}(x, y, z)$

A vector field associates a vector (or arrow) to every point in a space. Each vector represents both a magnitude and a direction acting at that point and time. Examples of vector fields include fluid velocity, vorticity, magnetic field, gravitational field.

$$\nabla \cdot \vec{V} = div\,\vec{V} = \frac{\partial V_x}{\partial x} + \frac{\partial V_y}{\partial x} + \frac{\partial V_z}{\partial x}$$

DIVERGENCE

Positive divergence is associated with expansion or 'generation' of a quantity at that point. Negative divergence is associated with a compression or 'destruction' of a quantity at that point. For incompressible fluid flow, divergence is zero; $\nabla \cdot \vec{V} = 0$. Gauss's laws for electromagnetism are for example $\nabla \cdot E = 4\pi\rho$ and $\nabla \cdot B = 0$.

$$\nabla \times \vec{V} = curl\,\vec{V} = \begin{vmatrix} \hat{\imath} & \hat{\jmath} & \hat{k} \\ \partial/\partial x & \partial/\partial y & \partial/\partial z \\ V_x & V_y & V_z \end{vmatrix}$$

CURL

1.10. MATRICES

Notation

$$[A]_{2\times 2} = \begin{bmatrix} A_{1,1} & A_{1,2} \\ A_{2,1} & A_{2,2} \end{bmatrix}$$

> Matrices may be distinguished from scalars and vectors using either bold or uppercase font, square brackets or accented with double underline. In linear algebra, the standard notation uses capital Latin letters for matrices and lowercase for vectors.

$[A]([B][C]) = ([A][B])[C]$ ASSOCIATIVE PROPERTY

> On the left-hand side of this identity, B is applied on C, then A is applied on (BC).

$[A]([B] + [C]) = [A][B] + [A][C]$ DISTRIBUTIVE PROPERTY

$[A][B] \neq [B][A]$ NOT COMMUTATIVE

> In general, the order of matrix multiplication is important.

$[A][A]^{-1} = [I]$ IDENTITY MATRIX

$$[I]_{3\times 3} = \begin{bmatrix} 1 & 0 & 0 \\ 0 & 1 & 0 \\ 0 & 0 & 1 \end{bmatrix}$$

$$([A][B])^{-1} = [B]^{-1}[A]^{-1}$$

Scalar Multiplication

$$\lambda[A] = \begin{bmatrix} \lambda A_{1,1} & \lambda A_{1,2} \\ \lambda A_{2,1} & \lambda A_{2,2} \end{bmatrix}$$

> To multiply a matrix by a scalar, multiply each element by the scalar.

Matrix Multiplication

$$[A][B] = \begin{bmatrix} A_{1,1}B_{11} + A_{1,2}B_{2,1} & A_{1,1} \cdot B_{1,2} + A_{1,2}B_{2,2} \\ A_{2,1}B_{1,1} + A_{2,2}B_{2,1} & A_{2,1}B_{1,2} + A_{2,2}B_{2,2} \end{bmatrix}$$

> To multiply matrices, each row of the first matrix operates on each column of the second matrix. The number of columns in A must equal the number of rows in B.

Rotation Matrices

$$R(\theta) = \begin{bmatrix} \cos\theta & -\sin\theta \\ \sin\theta & \cos\theta \end{bmatrix}$$

2D ROTATION ABOUT ORIGIN

$$R_x(\psi) = \begin{bmatrix} 1 & 0 & 0 \\ 0 & \cos\psi & -\sin\psi \\ 0 & \sin\psi & \cos\psi \end{bmatrix}$$

3D ROTATION ABOUT x

$$R_y(\theta) = \begin{bmatrix} \cos\theta & 0 & \sin\theta \\ 0 & 1 & 0 \\ -\sin\theta & 0 & \cos\theta \end{bmatrix}$$

3D ROTATION ABOUT y

$$R_z(\phi) = \begin{bmatrix} \cos\phi & -\sin\phi & 0 \\ \sin\phi & \cos\phi & 0 \\ 0 & 0 & 1 \end{bmatrix}$$

3D ROTATION ABOUT z

Screw Matrix

$$\vec{c} = \vec{a} \times \vec{b} = \begin{bmatrix} 0 & -a_z & a_y \\ a_z & 0 & -a_x \\ -a_y & a_x & 0 \end{bmatrix} \begin{pmatrix} b_x \\ b_y \\ b_z \end{pmatrix} = \begin{pmatrix} a_y b_z - a_z b_y \\ a_z b_x - a_x b_z \\ a_x b_y - a_y b_x \end{pmatrix}$$

The screw matrix represents the operation $[\vec{a} \times]$.

Euler Angles

$$[R] = Z(\psi)X(\theta)Z(\phi)$$

$$[R]\begin{pmatrix} x \\ y \\ z \end{pmatrix} = \begin{pmatrix} x' \\ y' \\ z' \end{pmatrix}$$

Euler angles rotates the local coordinate system into any orientation.

1. The first rotation is by an angle ϕ about the z-axis.
2. The second rotation is by an angle $\theta \in [0, \pi]$ about the former x-axis (now x').
3. The third rotation is by an angle ψ about the former z-axis (now z').

$$X(\phi)Z(\theta)X(\psi) = \begin{bmatrix} c\phi c\psi - c\theta s\phi s\psi & c\psi s\phi + c\theta c\phi s\psi & s\psi s\theta \\ -s\psi c\phi - c\theta s\phi c\psi & -s\psi s\phi + c\theta c\phi c\psi & c\psi s\theta \\ s\theta s\phi & -s\theta c\phi & c\theta \end{bmatrix}$$

$c\theta = \cos(\theta)$
$c\phi = \cos(\phi)$
$s\psi = \sin(\psi)$ *etc.*

Determinant

$$\det([A]_{2\times 2}) = |[A]_{2\times 2}| = \begin{vmatrix} a & b \\ c & d \end{vmatrix} = ad - bc$$

$$|[A]_{3\times 3}| = A_{1,1}\begin{vmatrix} A_{2,2} & A_{2,3} \\ A_{3,2} & A_{3,3} \end{vmatrix} - A_{1,2}\begin{vmatrix} A_{2,1} & A_{2,3} \\ A_{3,1} & A_{3,3} \end{vmatrix} + A_{1,3}\begin{vmatrix} A_{2,1} & A_{2,2} \\ A_{3,1} & A_{3,2} \end{vmatrix}$$

$$= (A_{2,2}A_{3,3} - A_{2,3}A_{3,2})A_{1,1} + (A_{2,3}A_{3,1} - A_{2,1}A_{3,3})A_{1,2} + (A_{2,1}A_{3,2} - A_{2,2}A_{3,1})A_{1,3}$$

The determinant of a geometric transformation matrix gives its transformation scale factor.

The determinant of a diagonal square matrix is the product of the diagonals.

Bold uppercase symbols are typically used in print to represent a matrix, e.g. A, M, Ω. Square parentheses, or underlined symbols are typically used in handwritten calculations to represent a matrix e.g. $[M]$, $\underline{\Omega}$.

Eigenvectors and Eigenvalues

Characteristic Equation

$$\det([A] - \lambda[I]) = 0$$

If A is an $n \times n$ transformation matrix, the solutions of the characteristic equation are eigenvalues λ_i.each having a corresponding eigenvector, $\vec{v}$.

$$[A]\vec{v} = \lambda\vec{v}$$

Solve the linear system of equations to find general eigenvectors, $\vec{v}$.

$$([A] - \lambda[I])\vec{v} = 0$$

In vibration analysis (e.g. for a mass-spring system), each eigenvector describes a mode shape of vibration, and each corresponding eigenvalue gives the oscillation frequency of that mode.

Matrix Inverse

Note: If $det([A]) = 0$, the matrix is non-invertible.

Inverse of a 2×2 Matrix

$$[A]_{2\times 2} = \begin{bmatrix} a & b \\ c & d \end{bmatrix}$$

$$\Rightarrow [A]^{-1} = \frac{1}{det([A])}\begin{bmatrix} d & -b \\ -c & a \end{bmatrix} = \frac{1}{ad - bc}\begin{bmatrix} d & -b \\ -c & a \end{bmatrix}$$

Inverse of any Square Matrix

$$[A]^{-1} = \frac{1}{\det([A])} adj([A])$$

First calculate the determinant, then find the **cofactor matrix.**
Next transpose the cofactor matrix to get the **adjugate**.

Cofactor Matrix

$$C_{ij} = (-1)^{i+j} M_{ij}$$

M_{ij} is the determinant of the 2×2 matrix obtained from A by removing the ith row and the jth column.

Adjugate Matrix

$$Adj([A]) = [C]^T,\ i.e.\ adj([A])_{ij} = C_{ji}$$

$$Adj\left([A]_{3\times 3}\right) = \begin{bmatrix} +\begin{vmatrix} A_{2,1} & A_{2,3} \\ A_{3,1} & A_{3,3} \end{vmatrix} & -\begin{vmatrix} A_{1,2} & A_{1,3} \\ A_{3,2} & A_{3,3} \end{vmatrix} & +\begin{vmatrix} A_{2,1} & A_{2,2} \\ A_{3,1} & A_{3,2} \end{vmatrix} \\ -\begin{vmatrix} A_{2,1} & A_{2,3} \\ A_{3,1} & A_{3,3} \end{vmatrix} & +\begin{vmatrix} A_{1,1} & A_{1,3} \\ A_{3,1} & A_{3,3} \end{vmatrix} & -\begin{vmatrix} A_{1,1} & A_{1,2} \\ A_{3,1} & A_{3,2} \end{vmatrix} \\ +\begin{vmatrix} A_{2,1} & A_{2,2} \\ A_{3,1} & A_{3,2} \end{vmatrix} & -\begin{vmatrix} A_{1,1} & A_{1,2} \\ A_{3,1} & A_{3,2} \end{vmatrix} & +\begin{vmatrix} A_{1,1} & A_{1,2} \\ A_{2,1} & A_{2,2} \end{vmatrix} \end{bmatrix}$$

1.11. CALCULUS

Derivatives

Leibniz's Notation

$$y, \frac{dy}{dx}, \frac{d^2y}{dx^2}, \frac{d^3y}{dx^3}, \ldots, \frac{d^ny}{dx^n}$$

Lagrange's Notation

$$f(x), f'(x), f''(x), f^{(3)}(x), \ldots, f^{(n)}(x)$$

Newton's Notation

$$y, \dot{y}, \ddot{y}, \dddot{y} \ldots$$

Elementary Rules for Differentiation

$$\frac{du}{dx} = u', \ \frac{dv}{dx} = v'$$

$$\frac{d(uv)}{dx} = vu' + uv'$$ PRODUCT RULE

$$\frac{d}{dx}\left(\frac{u}{v}\right) = \frac{vu' - uv'}{v^2}$$ QUOTIENT RULE

$$y = u(v(x)) \Rightarrow \frac{dy}{dx} = \frac{dy}{du} \cdot \frac{du}{dx}$$ CHAIN RULE

Integrals

INTEGRATION BY PARTS

$$\int u\, dv = uv - \int v\, du$$

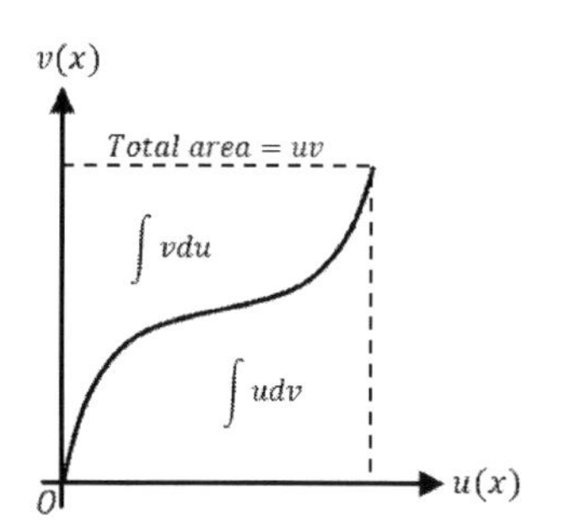

DEFINITE INTEGRATION BY PARTS

$$\int_a^b u(x) \cdot v'(x) dx = \left(\int_a^b u \cdot dv\right)$$

$$= [u(x)v(x)]_a^b - \int_a^b v(x)u'(x) dx$$

Table of Derivatives

$\dfrac{dy}{dx}$	y	$\int y\,dx$ (+const.)
ALGEBRAIC		
nx^{n-1}	x^n	$\dfrac{x^{n+1}}{n+1}, \quad n \neq -1$
$-x^{-2}$	$x^{-1} = \dfrac{1}{x}$	$ln\lvert x\rvert$
ae^{ax}	e^{ax}	$\dfrac{1}{a}e^{ax}, \quad a \neq 0$
$a^x \ln(a)$	a^x	$\dfrac{a^x}{\ln(a)}, \quad a > 0$
$x^{-1} = \dfrac{1}{x}$	$ln\, x$	$x\, ln\, x - x, \quad x > 0$
$\dfrac{f'(x)}{f(x)}$	$ln\, f(x)$	$x\, ln(f(x)) - \int x\, \dfrac{f'(x)}{f(x)} dx$
TRIGONOMETRIC		
$cos\, x$	$sin\, x$	$-cos\, x$
$-sin\, x$	$cos\, x$	$sin\, x$
$sec^2\, x$	$tan\, x$	$ln\lvert sec\, x\rvert$
$-cosec^2\, x$	$cot\, x$	$ln\lvert sin\, x\rvert$
$sec\, x\, tan\, x$	$sec\, x$	$ln\lvert sec\, x + tan\, x\rvert$
INVERSE TRIGONOMETRIC		
$\dfrac{1}{\sqrt{a^2 - x^2}}$	$arcsin \dfrac{x}{a}$	$x\, arcsin \dfrac{x}{a} + \sqrt{a^2 - x^2}$
$\dfrac{-1}{\sqrt{a^2 - x^2}}$	$arccos \dfrac{x}{a}$	$x\, arccos \dfrac{x}{a} - \sqrt{a^2 - x^2}$
$\dfrac{a}{a^2 + x^2}$	$arctan \dfrac{x}{a}$	$x\, arctan \dfrac{x}{a} - ln \sqrt{a^2 + x^2}$

$\frac{dy}{dx}$	y	$\int y\,dx\ (+const.)$
HYPERBOLIC		
$\cosh x$	$\sinh x$	$\cosh x$
$\sinh x$	$\cosh x$	$\sinh x$
$\operatorname{sech}^2 x$	$\tanh x$	$\ln(\cosh x)$
INVERSE HYPERBOLIC		
	$\frac{1}{\sqrt{x^2+a^2}}$	$\operatorname{arcsinh}\frac{x}{a}$
	$\frac{1}{\sqrt{x^2-a^2}}$	$\operatorname{arccosh}\frac{x}{a}$
	$\frac{a}{a^2-x^2}$	$\operatorname{arctanh}\frac{x}{a}$

Surface of Revolution

$$A = 2\pi \int_{x_1}^{x_2} y\sqrt{1+\left(\frac{dy}{dx}\right)^2}dx$$

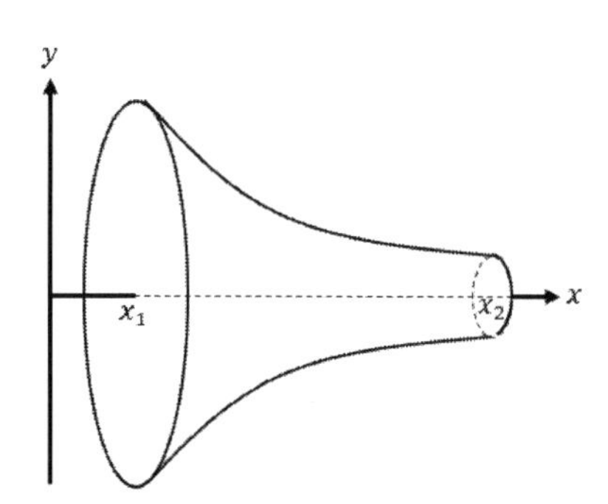

Volume of Revolution

$$V = \pi \int_{x_1}^{x_2} y^2 dx$$

Time Average of a Function

$$f_{avg} = \frac{1}{t_2 - t_1}\int_{t_1}^{t_2} f(t)dt$$

$$f_{rms} = \sqrt{\frac{1}{t_2 - t_1}\int_{t_1}^{t_2} [f(t)]^2 dt}$$

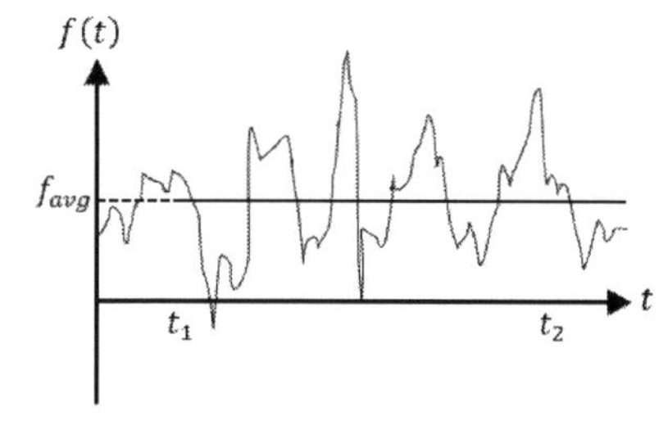

1.12. LAPLACE TRANSFORMS

For converting a function from the time domain $f(t)$ to the frequency domain $F(s)$; where (s) represents a variable in the complex plane, i.e. a complex number of the form $s = \sigma + j\omega$.

Definition

$$F(s) = L[f(t)] = \int_0^\infty e^{-st} f(t)dt, \ where \ Re(s) = 0$$

First order derivatives

$$L[y] = Y(s) \Rightarrow L\left[\frac{dy}{dt}\right] = sY(s) - y(0)$$

Second order derivatives

$$\Rightarrow L\left[\frac{d^2y}{dt^2}\right] = s^2Y(s) - sy(0) - \frac{dy}{dt}(0)$$

Laplace transforms can be used to solve Linear Ordinary Differential Equations given the initial conditions e.g. $y(0), y'(0)$ are known. The procedure is generally as follows:

1. Take the Laplace transform of the differential equation and plug in the initial conditions to convert it into a linear algebraic equation.
2. Rearrange for the output i.e. $X(s)$ or $Y(s)$, splitting into partial fractions if necessary.
3. Take the inverse Laplace transform of each term to find the solution to the original differential equation.

Table of Laplace Transforms

Function *for* $t > 0$	**Laplace Transform**
$f(t) = \mathcal{L}^{-1}\{F(s)\}$	$F(s) = \mathcal{L}\{f(t)\}$
$\delta(t)$ $(i.e. unit\ impulse\ at\ t = 0)$	1
$u(t)$ $(i.e. unit\ step\ at\ t = 0)$	$\frac{1}{s}$
a $(i.e. constant)$	$\frac{a}{s}$
e^{-at}	$\frac{1}{s+a}$
t	$\frac{1}{s^2}$
t^n $(where\ n\ is\ a\ positive\ integer)$	$\frac{n!}{s^{n+1}}$
te^{-at}	$\frac{1}{(s+a)^2}$
$\sin \omega t$	$\frac{\omega}{s^2+\omega^2}$
$\cos \omega t$	$\frac{s}{s^2+\omega^2}$
$\sin(\omega t + \theta)$	$\frac{s\sin(\theta) + \omega\cos(\theta)}{s^2+\omega^2}$
$\cos(\omega t + \theta)$	$\frac{s\cos(\theta) - \omega\sin(\theta)}{s^2+\omega^2}$

Function *for* $t > 0$	**Laplace Transform**
$f(t) = \mathcal{L}^{-1}\{F(s)\}$	$F(s) = \mathcal{L}\{f(t)\}$
$t\sin(\omega t)$	$\frac{2\omega s}{(s^2+\omega^2)^2}$
$t\cos(\omega t)$	$\frac{s^2-\omega^2}{(s^2+\omega^2)^2}$
$\sin(\omega t) - \omega t\cos(\omega t)$	$\frac{2\omega^3}{(s^2+\omega^2)^2}$
$\cos(\omega t) - \omega t\sin(\omega t)$	$\frac{s(s^2-\omega^2)}{(s^2+\omega^2)^2}$
$e^{-at}\sin\omega t$	$\frac{\omega}{(s+a)^2+\omega^2}$
$e^{-at}\cos\omega t$	$\frac{s+a}{(s+a)^2+\omega^2}$
$\frac{t}{2\omega}\sin\omega t$	$\frac{s}{(s^2+\omega^2)^2}$
$\frac{1}{2\omega^3}(\sin\omega t - \omega t\cos\omega t)$	$\frac{1}{(s^2+\omega^2)^2}$
$\frac{1}{2b^2}(\sin bt\sinh bt)$	$\frac{s}{s^4+4b^4}$
$\frac{1}{4b^3}(\sin bt\cosh bt - \cos bt\sinh bt)$	$\frac{1}{s^4+4b^4}$

1.13. STATISTICS

Discrete Random Variables

For a distribution X, consisting of a population of discrete random variables $[x_1, x_2, x_3, ...x_n]$:

$$E(X) = \mu = \frac{\Sigma x_i}{n}$$

$$var(X) = \sigma^2 = \frac{\Sigma(x_i - \bar{x})^2}{n}$$

$$\sigma = \sqrt{\frac{\Sigma(x_i - \bar{x})^2}{n}} = \sqrt{\frac{\sum x^2}{n} - \left(\frac{\sum x}{n}\right)^2}$$

> The sample mean is denoted $\bar{x}$ and the sample standard deviation is denoted s. For an estimate of population variance and standard deviation given just a sample of data, instead use $n - 1$ as the divisor in the equation above.

Grouped Frequency Distribution

$$E(X) = \bar{x} = \frac{\Sigma f x_i}{\Sigma f}$$

> For a grouped frequency distribution, use the mid-interval value as x.

$$var(X) = \sigma^2 = \frac{\sum f(x_i - \bar{x})^2}{n}$$

$$\sigma = \sqrt{\frac{\sum f(x_i - \bar{x})^2}{\sum f}} = \sqrt{\frac{\sum fx^2}{\sum f} - \left(\frac{\sum fx}{\sum f}\right)^2}$$

$E(x) = expected\ value\ (mean)$
$var(x) = population\ variance$
$\sigma = standard\ deviation$

Linear Regression (Least Squares Regression Line)

Method 1: Solution to Simultaneous Equations

The linear regression line minimises the sum of the residuals squared, i.e. $y_i = mx_i + c + \varepsilon_i$ minimises $\sum \varepsilon_i^2$ by solving the '*normal equations*':

$$\sum y_i = cN + m\sum x_i$$

$$\sum (x_i y_i) = c\sum x_i + m\sum x_i^2$$

For a given data set, first compute the sums, then next solve for m and c. These parameters create the least squares regression line in the form of $y = mx + c$.

Method 2: Gradient – Intercept

Find parameters for the equation $y = mx + c$:

$$m = \frac{\left(\sum y_i\right)\left(\sum x_i\right) - \left(\sum x_i\right)\left(\sum x_i y_i\right)}{N\left(\sum x_i^2\right) - \left(\sum x_i\right)^2}$$

$$c = \frac{N\left(\sum x_i y_i\right) - \left(\sum x_i\right)\left(\sum y_i\right)}{N\left(\sum x_i^2\right) - \left(\sum x_i\right)^2}$$

$N = number\ of\ data\ pairs\ (x_i, y_i)$

Pearson Product Moment Correlation Coefficient

The PMCC is a measure of the linear correlation between two variables X and Y.

$$r = \frac{\Sigma(x_i - \bar{x})\Sigma(y_i - \bar{y})}{\sqrt{\Sigma(x_i - \bar{x})^2}\sqrt{\Sigma(y_i - \bar{y})^2}}$$

Alternative method:

$$r = \frac{(N\Sigma xy) - \Sigma x\Sigma y}{\sqrt{[N\Sigma x^2 - (\Sigma x)^2][N\Sigma y^2 - (\Sigma y)^2]}}$$

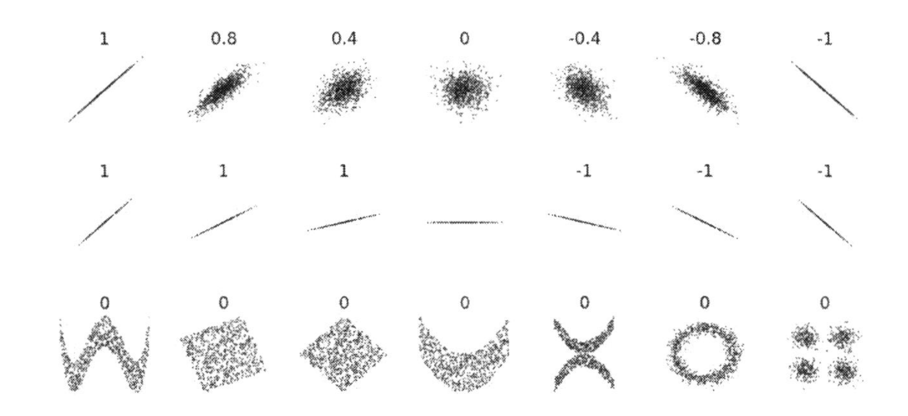

Spearman's Rank Correlation Coefficient

$$r_s = 1 - \frac{6\Sigma d_i^2}{n(n^2 - 1)}$$

$d_i = difference\ between\ the\ two\ ranks\ of\ each\ observation$
$n = number\ of\ observations$

Combinatorics

$$x! = x \cdot (x-1) \cdot (x-2) \cdot \ldots \cdot 3 \cdot 2 \cdot 1$$

FACTORIAL

$n!$ is the number of ways you can arrange n things. For example, 3! is $3 \cdot 2 \cdot 1 = 6$.
Example:
Consider three objects, labelled A B and C.
There are six orderings: ABC, ACB, BAC, BCA, CAB, CBA.

$$\binom{n}{k} = {}^nC_k = \frac{n!}{(n-k)!k!}$$

COMBINATIONS

The **Binomial Coefficient** $\binom{n}{k}$ or ${}^nC_k$ is the coefficient of the x^k term in the polynomial expansion of the binomial power $(1+x)^n$.

The Binomial Coefficient $\binom{n}{k}$ or or ${}^nC_k$ is read as 'n choose k' because there are $\binom{n}{k}$ ways to choose an unordered subset of k elements from a fixed set of n elements.
Example:
Consider four objects, labelled A, B, C and D.
${}^4C_2$ evaluates to 6.
There are six ways to choose two of them: AB, AC, AD, BC, BD, CD.

$${}^nP_k = \frac{n!}{(n-k)!}$$

PERMUTATIONS

There are in general more ways to choose objects from a set if you care about the order that they are chosen. This suggests that the device known as a "combination lock" is more accurately described as a *permutation* lock (the order is important).
Example:
Consider four objects, labelled A, B, C and D.
${}^4P_2$ evaluates to 12.
There are 12 permutations of two objects, where the order matters:
AB, AC, AD, BC, BD, CD, BA, CA, DA, CB, DB, DC.

Binomial Formula

$$(x+y)^n = \sum_{k=0}^{n} \binom{n}{k} x^{n-k} y^k = \binom{n}{0} x^n + \binom{n}{1} x^{n-1} y^1 + \binom{n}{2} x^{n-2} y^2 + \dots$$

$$+ \binom{n}{n-1} x y^{n-1} + \binom{n}{n} y^n$$

Pascal's Triangle

Row number n contains the numbers $\binom{n}{k}$ for $k = 0,1,2,\dots,n.$

Row															
0								1							
1							1		1						
2						1		2		1					
3					1		3		3		1				
4				1		4		6		4		1			
5			1		5		10		10		5		1		
6		1		6		15		20		15		6		1	
7	1		7		21		35		35		21		7		1

Pascal's Triangle allows the quick calculation of binomial coefficients without having to compute the coefficients.
Example: Using row 5 of the triangle above to expand $(x+y)^5$:
$(x+y)^5 = x^5 + 5x^4y + 10x^3y^2 + 10x^2y^3 + 5xy^4 + y^5$.

Binomial Distribution

$$X \sim B(n, p)$$

If $X \sim B(n,p)$ represents a discrete random variable with binomial distribution:
Each trial can result in just two possible outcomes. We might call one of these outcomes a success and the other, a failure. Each trial is independent, i.e. the outcome of one trial does not affect the outcome of any other trials.
n is the total number of repeated experiments.
p the probability of a single experiment yielding a successful result.
$q = 1 - p$ is the probability of a single experiment yielding a failure.

$$Mean = np$$

$$Variance = np(1-p) = npq$$

$$Median = \lfloor np \rfloor \ or \ \lceil np \rceil$$

$$Mode = \lfloor (n-1)p \rfloor \ or \ \lceil (n-1)p \rceil - 1$$

$\lfloor x \rfloor$ is the floor function (i.e. the greatest integer less than or equal to x.
$\lceil x \rceil$ is the ceiling function (i.e. the least integer greater than or equal to x.

Probability Mass Function (Binomial Distribution)

$$f(k, n, p) = P(X = k) = \binom{n}{k} p^k (1 - p)^{n-k}$$

The probability of getting exactly k successes in n trials with a binomially distributed random variable is given by the probability mass function.

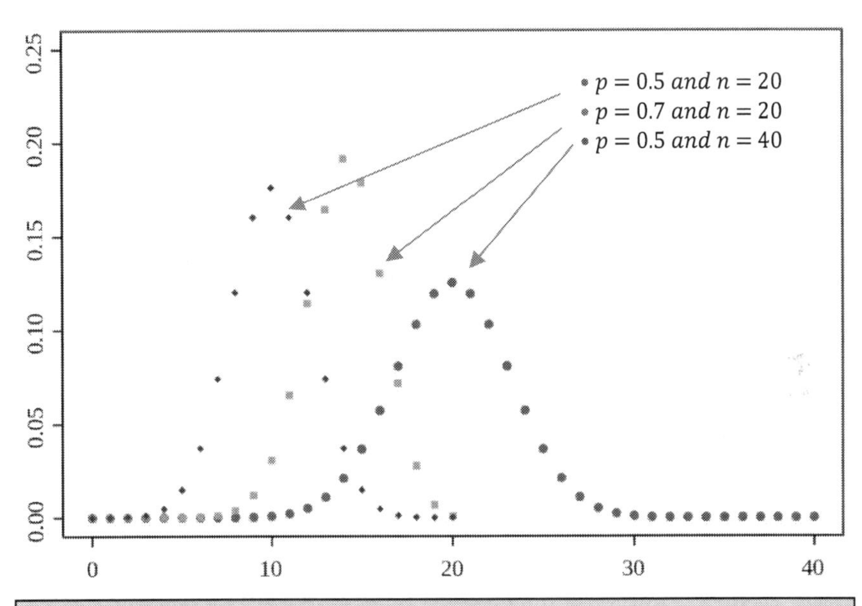

The horizontal axis above represents k successful outcomes.
The vertical axis represents the probability $P(X = k)$.

Cumulative Distribution Function (Binomial Distribution)

$$F_{binomial}(k; n, p) = P(X \leq k) = \sum_{i=0}^{\lfloor k \rfloor} \binom{n}{i} p^i (1 - p)^{n-i}$$

$\lfloor k \rfloor = \textit{the floor under } k \text{ (i.e. the greatest integer less than or equal to } k)$

Binomial Distribution

Cumulative Distribution Function

Probability of obtaining at most k successes in n independent trials, each of which has a probability p of success

$$F(k;\ n,p) = P(X \le k) = \sum_{i=0}^{\lfloor k \rfloor} \binom{n}{i} p^i (1-p)^{n-i}$$

	$p =$	0.01	0.05	0.10	0.15	0.20	0.25	0.30	0.35	0.40	0.45	0.50
$n = 2$	$k = 0$	0.9801	0.9025	0.8100	0.7225	0.6400	0.5625	0.4900	0.4225	0.3600	0.3025	0.2500
	$k = 1$	0.9999	0.9975	0.9900	0.9775	0.9600	0.9375	0.9100	0.8775	0.8400	0.7975	0.7500
	$k = 2$	1	1	1	1	1	1	1	1	1	1	1
$n = 3$	$k = 0$	0.9703	0.8574	0.729	0.6141	0.512	0.4219	0.343	0.2746	0.216	0.1664	0.125
	$k = 1$	0.9997	0.9928	0.972	0.9393	0.896	0.8438	0.784	0.7183	0.648	0.5748	0.5
	$k = 2$	1	0.9999	0.999	0.9966	0.992	0.9844	0.973	0.9571	0.936	0.9089	0.875
	$k = 3$	1	1	1	1	1	1	1	1	1	1	1
$n = 4$	$k = 0$	0.9606	0.8145	0.6561	0.522	0.4096	0.3164	0.2401	0.1785	0.1296	0.0915	0.0625
	$k = 1$	0.9994	0.986	0.9477	0.8905	0.8192	0.7383	0.6517	0.563	0.4752	0.391	0.3125
	$k = 2$	1	0.9995	0.9963	0.988	0.9728	0.9492	0.9163	0.8735	0.8208	0.7585	0.6875
	$k = 3$	1	1	0.9999	0.9995	0.9984	0.9961	0.9919	0.985	0.9744	0.959	0.9375
	$k = 4$	1	1	1	1	1	1	1	1	1	1	1
$n = 5$	$k = 0$	0.951	0.7738	0.5905	0.4437	0.3277	0.2373	0.1681	0.116	0.0778	0.0503	0.0313
	$k = 1$	0.999	0.9774	0.9185	0.8352	0.7373	0.6328	0.5282	0.4284	0.337	0.2562	0.1875
	$k = 2$	1	0.9988	0.9914	0.9734	0.9421	0.8965	0.8369	0.7648	0.6826	0.5931	0.5
	$k = 3$	1	1	0.9995	0.9978	0.9933	0.9844	0.9692	0.946	0.913	0.8688	0.8125
	$k = 4$	1	1	1	0.9999	0.9997	0.999	0.9976	0.9947	0.9898	0.9815	0.9688
	$k = 5$	1	1	1	1	1	1	1	1	1	1	1

Poisson Distribution

$$X \sim Poisson(\lambda)$$

If $X \sim Poisson(\lambda)$ represents a discrete random variable with Poisson distribution:
- Events are **rare** and occur at **random**.
- Events are **independent** of each other.
- The average number of events λ in the given interval is uniform and finite.
- The **average** number of successes (λ) that occurs in a specified region **is known**.
- The probability of an event in a small sub-interval is proportional to the length of the sub-interval.
- The probability that the event will occur in an extremely small region is virtually zero.

Examples
Misprinted letters in a novel, defective areas in several kilometres of cable, surface flaws on large sheets of aluminium, particles emitted by a radioactive source in a given time, car accidents on a stretch of highway, rare genetic mutations, shark attacks.

$$Mean = E(X) = \lambda$$

$$Variance = Var(X) = \lambda$$

$$Median \approx \left\lfloor \lambda + \frac{1}{3} - 0.02/\lambda \right\rfloor$$

$If\ \lambda\ is\ non-integer, the\ mode = \lfloor \lambda \rfloor$

$If\ \lambda\ is\ a\ positive\ integer, the\ modes\ are\ \lambda - 1, \lambda$

Probability Mass Function (Poisson Distribution)

$$f(\lambda,k) = P(X = k) = \frac{e^{-\lambda} \cdot \lambda^k}{k!}$$

The probability mass function gives the probability that the discrete random variable X (here obeying the rules of a Poisson distribution) occurs exactly k times in a given interval. Again, k is the number of times an event occurs in an interval and k can take values $0, 1, 2, ...$

If, instead of the average number of events, we are given a time rate r for the events to happen then in an interval of time t the expected value is $\lambda = rt$. Thus the probability of k events in time t is:

$$P(X = k) = e^{-rt} \frac{(rt)^k}{k!}$$

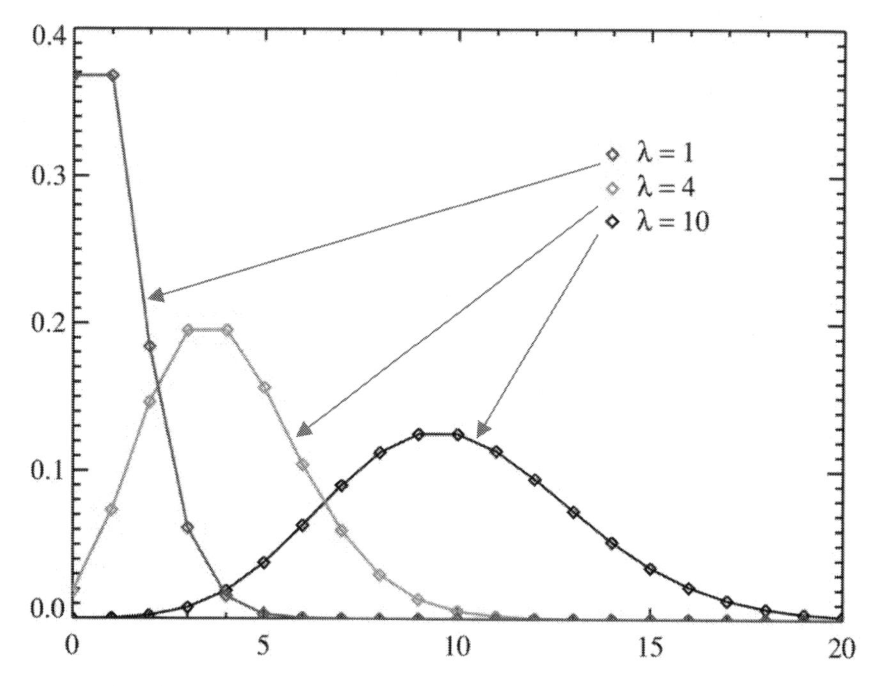

Cumulative Distribution Function (Poisson Distribution)

$$F_{poisson}(k;\lambda) = P(X \leq k) = e^{-\lambda}\sum_{i=0}^{\lfloor k \rfloor}\frac{\lambda^i}{i!}$$

$\lfloor k \rfloor = the\ floor\ under\ k\ \ (i.e.\ the\ greatest\ integer\ less\ than\ or\ equal\ to\ k)$

Poisson Distribution as a Binomial Approximation

> If you have a binomial distribution (e.g. yes/no or pass/fail), n is large (say >50) and p is small (say <0.1) then $B(n,p)$ can be approximated with $Poisson(\lambda)$ where $\lambda = np$.

$$X \sim B(n,p) \approx Poisson(np)$$

$$P(X = k) = e^{-np}\frac{(np)^k}{k!}$$

Poisson Distribution

Cumulative Distribution Function

Probability of obtaining at most x successes in n independent trials, each of which has a probability p of success

$$F(k;\lambda) = P(X \le k) = e^{-\lambda} \sum_{i=0}^{\lfloor k \rfloor} \frac{\lambda^i}{i!}$$

$\lambda =$	**0.001**	**0.005**	**0.010**	**0.015**	**0.020**	**0.025**	**0.030**	**0.035**	**0.040**	**0.045**	**0.050**
$k=0$	0.95123	0.99900	0.99501	0.99005	0.98511	0.97531	0.97045	0.96561	0.96079	0.95600	0.95123
$k=1$	0.99879	1	0.99999	0.99995	0.99989	0.99969	0.99956	0.99940	0.99922	0.99902	0.99879
$k=2$	0.99998	1	1	1	1	1	1	0.99999	0.99999	0.99999	0.99998
$k=3$	1	1	1	1	1	1	1	1	1	1	1

$\lambda =$	**0.060**	**0.070**	**0.080**	**0.095**	**0.100**	**0.200**	**0.300**	**0.400**	**0.500**	**0.600**	**0.700**
$k=0$	0.941765	0.932394	0.92312	0.90937	0.90484	0.81873	0.74082	0.67032	0.60653	0.54881	0.49659
$k=1$	0.998270	0.997661	0.99697	0.99576	0.99532	0.98248	0.96306	0.93845	0.90980	0.87810	0.84420
$k=2$	0.999966	0.999946	0.99992	0.99987	0.99985	0.99885	0.99640	0.99207	0.98561	0.97688	0.96586
$k=3$	1	1	1	1	1	0.99994	0.99973	0.99922	0.99825	0.99664	0.99425

$\lambda =$	**0.800**	**0.900**	**1.000**	**1.200**	**1.400**	**1.600**	**1.800**	**2.000**	**2.500**	**3.000**	**4.000**
$k=0$	0.44933	0.40657	0.36788	0.30119	0.24660	0.20190	0.16530	0.13534	0.08208	0.04979	0.01832
$k=1$	0.80879	0.77248	0.73576	0.66263	0.59183	0.52493	0.46284	0.40601	0.28730	0.19915	0.09158
$k=2$	0.95258	0.93714	0.91970	0.87949	0.83350	0.78336	0.73062	0.67668	0.54381	0.42319	0.23810
$k=3$	0.99092	0.98654	0.98101	0.96623	0.94627	0.92119	0.89129	0.85712	0.75758	0.64723	0.43347
$k=4$	0.99859	0.99766	0.99634	0.99225	0.98575	0.97632	0.96359	0.94735	0.89118	0.81526	0.62884
$k=5$	0.99982	0.99966	0.99941	0.99850	0.99680	0.99396	0.98962	0.98344	0.95798	0.91608	0.78513
$k=6$	0.99998	0.99996	0.99992	0.99975	0.99938	0.99866	0.99743	0.99547	0.98581	0.96649	0.88933
$k=7$	1	1	0.99999	0.99996	0.99989	0.99974	0.99944	0.99890	0.99575	0.98810	0.94887
$k=8$	1	1	1	1	0.99998	0.99995	0.99989	0.99976	0.99886	0.99620	0.97864
$k=9$	1	1	1	1	1	0.99999	0.99998	0.99995	0.99972	0.99890	0.99187
$k=10$	1	1	1	1	1	1	1	0.99999	0.99994	0.99971	0.99716

Normal Distribution

$$X \sim N(\mu, \sigma^2)$$

$X \sim N(\mu,\sigma^2)$ represents a Normal (i.e. Gaussian) Distribution or 'bell curve'.

$$Mean = median = mode = \mu$$

$$Variance = \sigma^2$$

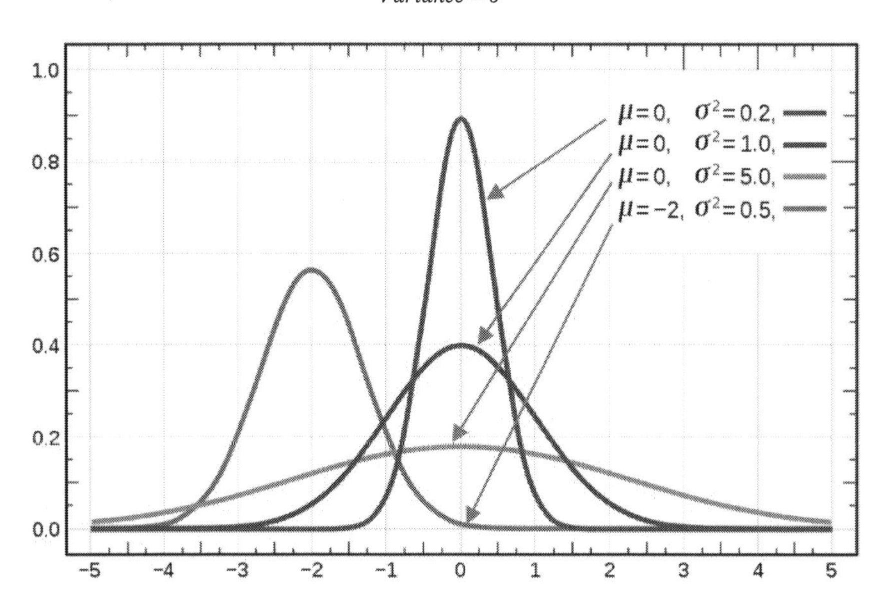

About 68% of values drawn from a normal distribution are within one standard deviation σ away from the mean; about 95% of the values lie within two standard deviations; and about 99.7% are within three standard deviations. This fact is known as the 68-95-99.7 (empirical) rule, or the 3-sigma rule. In manufacturing, 6 sigma referred originally to a process controlled to a 99.99966% success rate, or 3.4 defects per million opportunities (now it refers more broadly to a data-driven approach or set of tools and methodologies for eliminating defects.

Probability Density Function (Normal Distribution)

$$f(x \mid \mu, \sigma^2) = \frac{1}{\sqrt{2\pi\sigma^2}} e^{-\frac{(x-\mu)^2}{2\sigma^2}}$$

Standard Normal Distribution

Normalise values of X by using $z = \frac{X-\mu}{\sigma}$

This tables gives the probability that a normally distributed random variable z is equal to or smaller than z_1.

$Mean = 0$ $Standard\ deviation = 1$

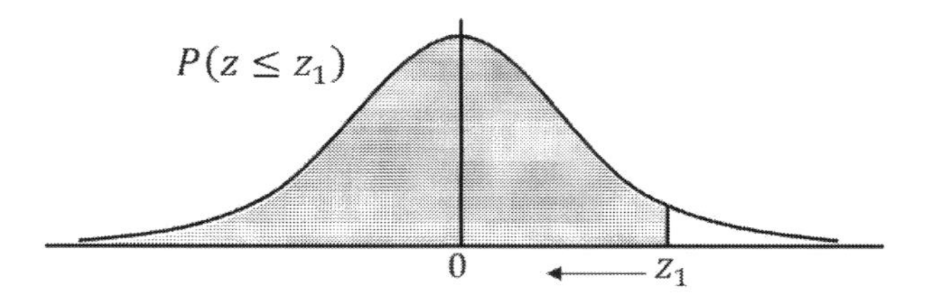

Z	.00	.01	.02	.03	.04	.05	.06	0.7	.08	.09
0.0	0.50000	0.50399	0.50798	0.51197	0.51595	0.51994	0.52392	0.52790	0.53188	0.53586
0.1	0.53983	0.54380	0.54776	0.55172	0.55567	0.55962	0.56356	0.56749	0.57142	0.57535
0.2	0.57926	0.58317	0.58706	0.59095	0.59483	0.59871	0.60257	0.60642	0.61026	0.61409
0.3	0.61791	0.62172	0.62552	0.62930	0.63307	0.63683	0.64058	0.64431	0.64803	0.65173
0.4	0.65542	0.65910	0.66276	0.66640	0.67003	0.67364	0.67724	0.68082	0.68439	0.68793
0.5	0.69146	0.69497	0.69847	0.70194	0.70540	0.70884	0.71226	0.71566	0.71904	0.72240
0.6	0.72575	0.72907	0.73237	0.73565	0.73891	0.74215	0.74537	0.74857	0.75175	0.75490
0.7	0.75804	0.76115	0.76424	0.76730	0.77035	0.77337	0.77637	0.77935	0.78230	0.78524
0.8	0.78814	0.79103	0.79389	0.79673	0.79955	0.80234	0.80511	0.80785	0.81057	0.81327
0.9	0.81594	0.81859	0.82121	0.82381	0.82639	0.82894	0.83147	0.83398	0.83646	0.83891

Standard Normal Distribution

$P(z \leq z_1)$; $z = \frac{X - \mu}{\sigma}$

Z	.00	.01	.02	.03	.04	.05	.06	0.7	.08	.09
1.0	0.84134	0.84375	0.84614	0.84849	0.85083	0.85314	0.85543	0.85769	0.85993	0.86214
1.1	0.86433	0.86650	0.86864	0.87076	0.87286	0.87493	0.87698	0.87900	0.88100	0.88298
1.2	0.88493	0.88686	0.88877	0.89065	0.89251	0.89435	0.89617	0.89796	0.89973	0.90147
1.3	0.90320	0.90490	0.90658	0.90824	0.90988	0.91149	0.91309	0.91466	0.91621	0.91774
1.4	0.91924	0.92073	0.92220	0.92364	0.92507	0.92647	0.92785	0.92922	0.93056	0.93189
1.5	0.93319	0.93448	0.93574	0.93699	0.93822	0.93943	0.94062	0.94179	0.94295	0.94408
1.6	0.94520	0.94630	0.94738	0.94845	0.94950	0.95053	0.95154	0.95254	0.95352	0.95449
1.7	0.95543	0.95637	0.95728	0.95818	0.95907	0.95994	0.96080	0.96164	0.96246	0.96327
1.8	0.96407	0.96485	0.96562	0.96638	0.96712	0.96784	0.96856	0.96926	0.96995	0.97062
1.9	0.97128	0.97193	0.97257	0.97320	0.97381	0.97441	0.97500	0.97558	0.97615	0.97670
2.0	0.97725	0.97778	0.97831	0.97882	0.97932	0.97982	0.98030	0.98077	0.98124	0.98169
2.1	0.98214	0.98257	0.98300	0.98341	0.98382	0.98422	0.98461	0.98500	0.98537	0.98574
2.2	0.98610	0.98645	0.98679	0.98713	0.98745	0.98778	0.98809	0.98840	0.98870	0.98899
2.3	0.98928	0.98956	0.98983	0.99010	0.99036	0.99061	0.99086	0.99111	0.99134	0.99158
2.4	0.99180	0.99202	0.99224	0.99245	0.99266	0.99286	0.99305	0.99324	0.99343	0.99361

Standard Normal Distribution

$P(z \leq z_1)$; $z = \frac{X - \mu}{\sigma}$

z	.00	.01	.02	.03	.04	.05	.06	0.7	.08	.09
2.5	0.99379	0.99396	0.99413	0.99430	0.99446	0.99461	0.99477	0.99492	0.99506	0.99520
2.6	0.99534	0.99547	0.99560	0.99573	0.99585	0.99598	0.99609	0.99621	0.99632	0.99643
2.7	0.99653	0.99664	0.99674	0.99683	0.99693	0.99702	0.99711	0.99720	0.99728	0.99736
2.8	0.99744	0.99752	0.99760	0.99767	0.99774	0.99781	0.99788	0.99795	0.99801	0.99807
2.9	0.99813	0.99819	0.99825	0.99831	0.99836	0.99841	0.99846	0.99851	0.99856	0.99861
3.0	0.99865	0.99869	0.99874	0.99878	0.99882	0.99886	0.99889	0.99893	0.99896	0.99900
3.1	0.99903	0.99906	0.99910	0.99913	0.99916	0.99918	0.99921	0.99924	0.99926	0.99929
3.2	0.99931	0.99934	0.99936	0.99938	0.99940	0.99942	0.99944	0.99946	0.99948	0.99950
3.3	0.99952	0.99953	0.99955	0.99957	0.99958	0.99960	0.99961	0.99962	0.99964	0.99965
3.4	0.99966	0.99968	0.99969	0.99970	0.99971	0.99972	0.99973	0.99974	0.99975	0.99976
3.5	0.99977	0.99978	0.99978	0.99979	0.99980	0.99981	0.99981	0.99982	0.99983	0.99983
3.6	0.99984	0.99985	0.99985	0.99986	0.99986	0.99987	0.99987	0.99988	0.99988	0.99989
3.7	0.99989	0.99990	0.99990	0.99990	0.99991	0.99991	0.99992	0.99992	0.99992	0.99992
3.8	0.99993	0.99993	0.99993	0.99994	0.99994	0.99994	0.99994	0.99995	0.99995	0.99995
3.9	0.99995	0.99995	0.99996	0.99996	0.99996	0.99996	0.99996	0.99996	0.99997	0.99997
4.0	0.99534	0.99547	0.99560	0.99573	0.99585	0.99598	0.99609	0.99621	0.99632	0.99643

2. MATERIALS

2.1. THE ELEMENTS

Material	Atomic Number	Density [kg m^{-3}]	Melting Point [°C]	Boiling Point [°C]	Crystal Structure
Actinium	89	10100	1050	3200	fcc
Aluminium	13	2700	660	2470	fcc
Americium	95	-	-	-	-
Antimony	51	6700	630	1380	rho
Argon	18	1.66	-190	-186	fcc
Arsenic	33	5730	820	613	rho
Astatine	85	-	250	350	-
Barium	56	3600	730	1640	bcc
Berkelium	97	-	-	-	-
Beryllium	4	1800	1280	2970	hcp/cub
Bismuth	83	9800	271	1560	rho
Bohrium	107	-	-	-	-
Boron	5	2500	2330	2550	tetra
Bromine	35	3100	-7.3	59	ortho
Cadmium	48	8650	321	765	hcp
Caesium	55	1870	28.4	690	bcc
Calcium	20	1540	850	1490	fcc/bcc
Californium	98	-	-	-	-
Carbon	6	2300	>2500	4830	h/c/d
Cerium	58	6800	800	3470	fcc/h/f/b
Chlorine	17	3.21	-101	-34.7	tetra
Chromium	24	7200	1890	2482	bcc
Cobalt	27	8900	1490	2900	hcp/fcc
Copernicium	112	-	-	-	-
Copper	29	8930	1080	2600	fcc
Curium	96	-	-	-	-
Darmstadtium	110	-	-	-	-
Dubnium	105	-	-	-	-

Material	Atomic Number	Density [kg m^{-3}]	Melting Point [°C]	Boiling Point [°C]	Crystal Structure
Dysprosium	66	8500	1410	2630	rho/hcp
Einsteinium	99	-		-	
Erbium	68	9000	1497	2927	hcp
Europium	63	5200	826.8	1440	bcc
Fermium	100	-	-	-	-
Flerovium	114	-	-	-	-
Fluorine	9	1.7	-220	-188	-
Francium	87	-	29.8	647	-
Gadolinium	64	7900	1312	2727	hcp/bcc
Gallium	31	5950	30	2403	fc.orth
Germanium	32	5400	940	2827	cub (dia)
Gold	79	19300	1063	2966	fcc
Hafnium	72	13300	2150	5427	hcp/bcc
Hassium	108	-	-	-	-
Helium	2	0.166	-272.3	-269	hcp/cub
Holmium	67	8800	1461	2627	hcp
Hydrogen	1	0.08987	-259.2	-253	hcp/cub
Indium	49	7310	157	2027	bct
Iodine	53	4940	113.4	184	ortho
Iridium	77	22420	2443	4527	fcc
Iron	26	7870	1535	3027	bcc/f/bcc
Krypton	36	-153.2	-157.4	3469	fcc
Lanthanum	57	6150	920	3464	hep/f/bcc
Laurencium	103	-	-	-	-
Lead	82	11340	327	1744	fcc
Lithium	3	534	179	1317	hep/f/bcc
Livermorium	116	-	-	-	-
Lutetium	71	9800	1652	3327	hcp
Magnesium	12	1741	650	1107	hcp
Manganese	25	7440	1244	2097	cub
Meitnerium	109	-	-	-	-
Mendelevium	101	-	-	-	-
Mercury	80	13590	-39	356	rho

Material	**Atomic Number**	**Density** [kg m^{-3}]	**Melting Point** [°C]	**Boiling Point** [°C]	**Crystal Structure**
Molybdenum	42	10200	2607	5557	bcc
Neodymium	60	6960	1024	3027	hcp/bcc
Neon	10	0.839	-249	-246	fcc
Neptunium	93	-	-	-	-
Nickel	28	8900	1453	2732	fcc
Niobium	41	8.57	2468	4927	bcc
Nitrogen	7	1.165	-209.7	-196	cub/hcp
Nobelium	102	-	-	-	-
Osmium	76	22480	3027	4630	hcp
Oxygen	8	1.33	-218.3	-183	rhom
Palladium	46	12000	1552	2930	fcc
Phosphorus	15	2200	44.2	280	cub
Platinum	78	21450	1770	3830	fcc
Plutonium	94	-	-	-	-
Polonium	84	9400	254	962	monoc
Potassium	19	860	63.8	774	bcc
Praseodymium	59	6800	935	3130	hcp/bcc
Promethium	61	-	1035	2730	-
Protactinium	91	15400	1230	4030	tetra
Radium	88	5000	700	1140	-
Radon	86	9.73	-71	-61.7	-
Rhenium	75	20500	3180	5630	hcp
Rhodium	45	12440	1960	3730	fcc
Roentgenium	111	-	-	-	
Rubidium	37	1530	39	688	bcc
Ruthenium	44	12400	2250	3930	hcp
Rutherfordium	104	-	-	-	-
Samarium	62	7500	1072	1930	rho/b
Scandium	21	3000	1540	2730	hcp/fcc
Seaborgium	106	-	-	-	-
Selenium	34	4810	217	685	hcp
Silicon	14	2300	1407	2355	cub
Silver	47	10500	961	2212	fcc/hcp

Material	Atomic Number	Density [kg m^{-3}]	Melting Point [°C]	Boiling Point [°C]	Crystal Structure
Sodium	11	970	98	892	bcc
Strontium	38	2600	769	1384	fcc/h/b
Sulphur	16	2070	113	445	fc orth
Tantalum	73	16600	3000	5425	bcc
Technetium	43	11400	2230	4630	hcp
Tellurium	52	6240	450	990	hcp
Terbium	65	8300	1356	2830	hcp/rho
Thallium	81	11860	303.6	1460	hcp/fcc
Thorium	90	11500	1727	4230	fcc/bcc
Thulium	69	9300	1545	1730	hcp/bcc
Tin	50	7300	232	2270	cub/bcc
Titanium	22	4540	1675	3260	hcp/bcc
Tungsten	74	19320	3380	5930	bcc
Oganesson	118	-	-	-	-
Moscovium	115	-	-	-	-
Tennessine	117	-	-	-	-
Nihonium	113	-	-	-	-
Uranium	92	19050	1132	3820	rho/tetra
Vanadium	23	6100	1890	3030	bcc
Xenon	54	5.5	-112	-107	fcc
Ytterbium	70	7000	824	1430	fcc/bcc
Yttrium	39	4600	1495	2930	hcp/bcc
Zinc	30	7140	420	910	hcp
Zirconium	40	6500	1850	3580	hcp/bcc

Crystal Structure
bcc: body-centred cubic.
fcc: face-centred cubic.
hcp: hexagonal close-packed.
cub: cubic.
bct: body-centred tetragonal.
h/c/g: (carbon) hexagonal / cubic diamond / 2 interconnected fcc lattices.
tetra: tetragonal.
monoc: monoclinic.
orth: orthorhombic.
fc orth: face-centred orthorhombic.

Periodic Table

	1	2		3	4	5	6	7	8	9	10	11	12	13	14	15	16	17	18
1	1 H																		2 He
2	3 Li	4 Be												5 B	6 C	7 N	8 O	9 F	10 Ne
3	11 Na	12 Mg												13 Al	14 Si	15 P	16 S	17 Cl	18 Ar
4	19 K	20 Ca		21 Sc	22 Ti	23 V	24 Cr	25 Mn	26 Fe	27 Co	28 Ni	29 Cu	30 Zn	31 Ga	32 Ge	33 As	34 Se	35 Br	36 Kr
5	37 Rb	38 Sr		39 Y	40 Zr	41 Nb	42 Mo	43 Tc	44 Ru	45 Rh	46 Pd	47 Ag	48 Cd	49 In	50 Sn	51 Sb	52 Te	53 I	54 Xe
6	55 Cs	56 Ba	*	71 Lu	72 Hf	73 Ta	74 W	75 Re	76 Os	77 Ir	78 Pt	79 Au	80 Hg	81 Tl	82 Pb	83 Bi	84 Po	85 At	86 Rn
7	87 Fr	88 Ra	**	103 Lr	104 Rf	105 Db	106 Sg	107 Bh	108 Hs	109 Mt	110 Ds	111 Rg	112 Cn	113 Nh	114 Fl	115 Mc	116 Lv	117 Ts	118 Og
			*	57 La	58 Ce	59 Pr	60 Nd	61 Pm	62 Sm	63 Eu	64 Gd	65 Tb	66 Dy	67 Ho	68 Er	69 Tm	70 Yb		
			**	89 Ac	90 Th	91 Pa	92 U	93 Np	94 Pu	95 Am	96 Cm	97 Bk	98 Cf	99 Es	100 Fm	101 Md	102 No		

Pure Metallic Solids – Mechanical

Material	Density [kg m⁻³]	Melting Point [°C]	UTS [MPa]	Yield Strength [MPa]	Young's Modulus [GPa]	Poisson's Ratio
	ρ	T_m	c_p	k	α	ν
Aluminium[2,3]	2700	660	80	50	71	0.34
Antimony	6700	630	11	-	78	-
Barium	3600	730	-	-	-	-
Beryllium	1800	1280	483	345	296	0.05
Bismuth	9800	271	-	-	32	-
Cadmium	8650	321	71	-	60	0.33
Chromium[2]	7200	1890	413	362	0.25	-
Cobalt	8900	1490	944	758	211	0.32
Copper[2,3]	8930	1080	150	75	117	0.34
Gallium	5950	30	-	-	-	-
Germanium	5400	940	-	-	128	-
Gold	19300	1063	103	-	78	0.44
Hafnium[2]	13300	2150	-	-	138	-
Indium	7310	157	2.6	-	11	-
Iridium	22420	2443	1100	-	517	0.26
Iron (pure)[2,3]	7870	1535	300	165	208	0.29
Lead	11340	327	15	12	18	0.43
Magnesium[2,3]	1741	650	190	95	44	0.29
Manganese	7440	1244	496	241	191	0.35
Molybdenum	10200	2607	500	-	290	0.31
Nickel[2,3]	8900	1453	300	60	207	0.31
Niobium	8.57	2468	585	207	103	0.38
Osmium	22480	3027	-	-	558	-
Platinum	21450	1770	350	-	150	0.38
Rhodium	12440	1960	951	-	283	-
Silicon	2300	1407	-93	-	113	-

Pure Metallic Solids – Mechanical

Material	**Density** [kg m^{-3}]	**Melting Point** [°C]	**UTS** [MPa]	**Yield Strength** [MPa]	**Young's Modulus** [GPa]	**Poisson's Ratio**
	ρ	T_m	c_p	k	α	ν
Silver	10500	961	125	-	83	0.37
Strontium	2600	769	-	-	15.7	-
Tantalum	16600	3000	350	200	186	0.35
Tin	7300	232	30	-	42	0.33
Titanium[2]	4540	1675	235	140	107	0.36
Tungsten	19320	3380	350	100	408	0.28
Uranium	19050	1132	400	200	190	-
Vanadium	6100	1890	~600	~550	130	0.36
Zinc	7140	420	150	-	110	0.25
Zirconium[2]	6500	1850	500	200	99	0.35

1 Properties are temperature dependent. Properties at Standard Conditions used where possible (20°C, 101325 Pa).

2 Properties will be significantly enhanced by alloying.

3 See following section for tables on alloys.

UTS Ultimate Tensile Strength

Pure Metallic Solids – Thermal and Electrical

Material	Melting Point [°C]	Specific heat capacity [J kg^{-1} K^{-1}]	Thermal Conductivity [W m^{-1} K^{-1}]	Coefficient of Linear Expansion [μm m^{-1} K^{-1}]	Electrical Resistivity [nΩ m]	Temperature Coefficient of Resistance [×10^{-3} K^{-1}]
	T_m	c_p	k	α	ρ	α
Aluminium[2,3]	660	913	237	23	26.5	4.29
Antimony	630	207	25.9	10	370	4.0
Barium	730	67	-	18	500	-
Beryllium	1280	1886	190	12	~50	6.0
Bismuth	271	122	8	13	1050	4.5
Cadmium	321	230	92.1	31	73	4.0
Chromium[2]	1890	460	66.9	6.2	129	5.88
Cobalt	1490	414	69	13.8	52.5	6.6
Copper[2,3]	1080	386	401	16.5	16.7	3.86
Gallium	30	372	33	11.5	150	6.04
Germanium	940	322	58.6	5.7	0.45Ωm	-
Gold	1063	129	318	14.2	20.1	3.4
Hafnium[2]	2150	147	20.9	5.9	350	3.8
Indium	157	233	86.6	32.1	80	4.7
Iridium	2443	130	147	6.8	47	3.9
Iron (pure)[2,3]	1535	106	90	112	100	6.2
Lead	327	126	35	26.5	210	4.3
Magnesium[2,3]	650	1025	418	25	44.5	4.3
Manganese	1244	477	-	23	1440	-
Molybdenum	2607	276	142	5.1	52	4.6
Nickel[2,3]	1453	471	82.9	13.3	684	6.0
Niobium	2468	268	54.4	6.9	132	3.95
Osmium	3027	130	-	3.2	95	4.2
Platinum	1770	132	71.1	9.1	110	3.8
Rhodium	1960	247	150	8.3	45.1	4.57

Pure Metallic Solids – Thermal and Electrical

Material	**Melting Point** [°C]	**Specific heat capacity** [J kg^{-1} K^{-1}]	**Thermal Conductivity** [W m^{-1} K^{-1}]	**Coefficient of Linear Expansion** [μm m^{-1} K^{-1}]	**Electrical Resistivity** [nΩ m]	**Temperature Coefficient of Resistance** [×10^{-3} K^{-1}]
	T_m	c_p	k	α	ρ	α
Silicon	1407	678	83.7	2.8	100	-
Silver	961	235	428	19	14.7	1.0
Strontium	769	-	-	-	227.6	-
Tantalum	3000	142	54.4	6.5	135	3.8
Tin	232	222	63	28	0.11	5.0
Titanium[2]	1675	522	11.4	8.41	420	3.8
Tungsten	3380	138	180	4.5	53	4.6
Uranium	1132	117	27.6	varies	300	3.4
Vanadium	1890	498	31	8.3	250	2.8
Zinc	420	382	113	31	58.9	4.2
Zirconium[2]	1850	289	21.1	5.85	450	4.4

1 Properties are temperature dependent. Properties at *Standard Conditions* used where possible (20°C, 101325 Pa).

2 Properties will be significantly altered by alloying. See Section 2.2 and 2.3 for Steels and Alloys.

Caution: Coefficient of linear expansion and temperature coefficient of resistance both use the same symbol, α.

2.2. STEELS AND ALLOYS

IRON AND STEEL

Properties of selected ferrous metals

Material	BS Grade (% Carbon content)	UTS [MPa]	Yield Strength [MPa]	BHN	Young's Modulus E [MPa]	Application Notes
Grey Iron BS 1452	10 (1.5–4.3)	160	-620	160-180	76-104	Brittle, low tensile strength, high compressive strength, easy to cast, can machine to a good polished surface finish
Grey Iron BS 1452	24 (1.5–4.3)	370	-1240	240-300	124-145	
Spheroidal Graphite BS2789	SNG37/2	570		210-310		Nodular iron, nearly as good as steel. Good ductility. Also known as Ductile Cast Iron, Nodular Cast Iron, Spheroidal Graphite Iron and Spherulitic Cast Iron
Malleable White Heart BS 309	W22/24	340	+200	248	170	Good casting properties. Better ductility than gray cast irons, good tensile strength.

Yield strength in tension (+) and compression (-)

UTS Ultimate Tensile Strength

BHN Brinell Hardness Number

Poisson's Ratios: Cast Iron $\nu \approx 0.211$.

Temperature coefficient of (electrical) resistance of iron is approximately 0.00651.

CARBON STEELS

Material	BS970	AISI equivalent	Composition	Ultimate Tensile Strength [MPa]	Yield strength [MPa]	Brinell Hardness BHN	Application Notes
Dead mild	070 M20	1020	0.20 C, 0.7 Mn	400	200	125-180	Easy machinable, weldable, light stresses, low strength
Mild	070 M26	1026	0.26 C, 0.7 Mn	430	215	140-190	Stronger than M20, good machinability, weldable
Medium Carbon	080 M30	1030	0.30 C, 0.8 Si	460	230	140-190	Tough, for forgings, nuts, bolts, spanners, Hardened; use up to 20mm section. Slightly less machinable
(with typical heat treatment)				550-700*	340	150-210	
Medium Carbon	080 M46	1043	0.46 C, 0.8 Mn	460	280	150-210	Motor shafts, axles, brackets, couplings
Medium Carbon	080 M50	1050	0.50 C, 0.8 Mn	570	280	180-230	Structural steel. Used in gears, shafts, axles, bolts, studs, and machine tool parts.
(with typical heat treatment)				700-850*	430	200-255	
Carbon Manganese	216 M28	1137	0.28 C, 0.25 Si, 1.3 Mn	540	400	150-210	Strength and toughness due to high Mn.
Case Hardening	080 M15	1065	0.15 C, 0.25 Si, 0.8 Mn	460	300		Used where wear is important: gears, pawls
Spring	060 A96	1090	0.96 C, 0.50 Si, 0.6 Mn	1300		500	Springs, knives, taps, dies, milling cutters

Young's Modulus = 210 GPa. Poisson's Ratio = 0.292

ALLOY STEELS

Material	BS970	AISI equivalent	Composition	Ultimate Tensile Strength [MPa]	Fatigue limit [MPa]	Corrosion resistance	Weldability	Application Notes
Low Alloy Steel								
Structural steel	709 M40	4140	1.0 Cr, Mo	900	400	poor	†	Structures, high tensile shafts, etc.
Nickel/ Chrome/Mo	835 M30		4.25 Ni, Cr, Mo	1550	700	Poor	†	For high strength at elevated temperatures
Chrome/ Mo/ Vanadium	897 M39		3.0 Cr, Mo, V	1300	620	poor	†	Used for high temp applications
Stainless Steel								
Martensitic (hardened)	431 S29		15-20 Cr, 2-3 Ni	880	340	good	poor	Resistant to corrosion and tempering at high temp
High Tensile Steel								
Nickel/ Chrome/Mo	817 M40	970	1.7 Ni, 0.4 Cr	1540	-	good	good	Direct hardening, for dies and shear blades
Austenitic	301 S21	301	13 Cr, Ni, Mo	540-1200 *	260	good	good	Stainless and heat resisting
Maraging	310 S31	310	25 Cr, 20 Ni	1800	-	good	good	Corrosion/ wear resisting. Hard to m/c

Mo Molybdenum

*Depends on the amount of tempering

†Weld area needs pre and post heat treatment

Poisson's Ratios: Steel ≈ 0.27-0.30. Ni-Steel ≈ 0.291. Stainless ≈ 0.305.

2.3. ALLOYS

ALLOYS – Mechanical

Material	Alloy composition [%]	Density [kg m^{-3}]	Melting Point [°C]	Ultimate Tensile Strength[1] [MPa]	Young's Modulus [GPa]	Application Notes
Aluminium Alloys						
Aluminium Copper	Cu 8	2830	-	440	73	Wrought is corrosion and oxidation resistant
Aluminium Silicon LM6	Si 11.5	2650	-	160-190	75	Castings for food, chemical and marine applications
Y Alloy	Cu 4	2780	-	380 (420)*	71	Strong, hard, heat treatable
Aluminium Alloy 2014A	Cu 4, Mg 1.5, Si 1	2800	-	440	71	General purpose alloy, wrought or forged
Aluminium Alloy 6061	Cu 4, Mg 4, Mn 0.6	2910	-	190	71	Extrusions for general engineering
Copper Alloys						
Arsenical Copper	As 0.35, P 0.024	8940	1083	220 (360)*	120	Strong at high temp. E.g. heat exchangers
Copper Zirconium	Zr 0.15	8940	1100	230 (495)*	120	High conductivity, high temperature uses
70/30 Brass	Zn 30	8520	954	270 (600)*	110	Ductile for deep drawing, presswork
60/40 Brass	Zn 40	8380	904	350	100	Condenser, heat exchanger plates
Aluminium Bronze	Al 8	7750	1041	430 (660)*	120	Imitation jewellery, condenser tubes
Bronze	Sn 10	8900	1280	280	75	Bearings, bushes, springs,
Phosphor Bronze	Sn 5, P 0.1	8920	1050	280	115	
Nickel Silver	Zn 27, Ni 10	8610	1010	375 (650)*	130	

ALLOYS – Mechanical

Material	Alloy composition [%]	Density [kg m^{-3}]	Melting Point [°C]	Ultimate Tensile Strength[1] [MPa]	Young's Modulus [GPa]	Application Notes
Nickel-Alu-Bronze	Al 10, Fe 5, Ni 5	7530	1038-1054	655	110	Ship's propeller blades, hub and bolts; resistant to erosion-corrosion, abrasive wear and cavitation
Cupro-Nickel	Ni 10, Fe 2, Mn 0.5	8890	1145	360 (600)*	140	False silver coinage, condenser and heat exchanger tubes, saltwater piping
Magnesium Alloy						
Magnesium Alloy	Al 6, Zn 1, Mn 0.5	1750	630-640	250	40	Sheet, tube, extruded, forged
Nickel-Chrome Alloys[2]						
Monel 400	Cu 31, Fe 2.5	8830	1300-1350	550-760	-	Strong, tough, ductile, corrosion resistant
Inconel 600	Cr 15.5, Fe 8	8420	1370-1425	629	-	Oxidation resistant, good high temp properties
Nimonic 75	Cr 20	8370	1340-1380	*420 at 400°C*	-	Resistant to creep, fatigue, oxidation, thermal shock
Nimonic 90	Cr 20, Co 17	8180	1310-1370	*420 at 400°C*	-	Strong at high temperatures. Aerospace applications.
Incoloy 800	Fe + Ni 32.5, Cr 21	7950	1355-1385	524	-	Tubes in furnaces, radiant heaters
Incoloy 825	Fe + Ni 42, Cr 21.5, Mo 3	8140	1370-1400	672	-	For corrosive conditions, resists stress corrosion

1 *Tensile strength x (y) means: "annealed (fully hardened)".

2 Nickel-alloys used over large temperature range. Properties vary with temperature.

ALLOYS – Thermal and Electrical

Material	Alloy composition [%]	Density [kg m^{-3}]	Melting Point [°C]	Thermal Conductivity [W m^{-1} K^{-1}]	Resistivity [nΩ m^{-1}]	Coefficient of Thermal Expansion [10^{-6} °C]
Aluminium Alloys						
Aluminium Copper	Cu 8	2830	-	218	47	22.5
Aluminium Silicon LM6	Si 11.5	2650	-	142	46	20.0
Y Alloy	Cu 4	2780	-	126	52	22.5
Aluminium Alloy 2014A	Cu 4, Mg 1.5, Si 1	2800	-	147	50	22.5
Aluminium Alloy 6061	Cu 4, Mg 4, Mn 0.6	2910	-	151	49	23.5
Copper Alloys						
Arsenical Copper	As 0.35, P 0.024	8940	1083	177	-	17.4
Copper Zirconium	Zr 0.15	8940	1100	195	-	-
70/30 Brass	Zn 30	8520	954	122	68.7	19.9
60/40 Brass	Zn 40	8380	904	127	68.1	20.8
Aluminium Bronze	Al 8	7750	1041	80	114	17.8
Bronze	Sn 10	8900	1280	46	180	19.0
Phosphor Bronze	Sn 5, P 0.1	8920	1050	75	95	18.0
Nickel Silver	Zn 27, Ni 10	8610	1010	37	-	16.4
Nickel-Alu-Bronze	Al 10, Fe 5, Ni 5	7530	1038-1054			
Cupro-Nickel	Ni 10, Fe 2, Mn 0.5	8890	1145	42	-	15.7
Magnesium Alloy						
Magnesium Alloy	Al 6, Zn 1, Mn 0.5	1750	630-640	117	60	26.5

ALLOYS – Thermal and Electrical

Material	Alloy composition [%]	Density [kg m^{-3}]	Melting Point [°C]	Thermal Conductivity [W m^{-1} K^{-1}]	Resistivity [nΩ m^{-1}]	Coefficient of Thermal Expansion [10^{-6} °C]
Nickel-Chrome Alloys[2]						
Monel 400	Cu 31, Fe 2.5	8830	1300-1350	22	51	14.1
Inconel 600	Cr 15.5, Fe 8	8420	1370-1425	15	103	13.3
Nimonic 75	Cr 20	8370	1340-1380	12	109	11.0
Nimonic 90	Cr 20, Co 16, Fe<5	8180	1310-1370	12	118	12.7
Incoloy 800	Fe + Ni 32.5, Cr 21	7950	1355-1385	12	99	14.2
Incoloy 825	Fe + Ni 42, Cr 21.5, Mo 3	8140	1370-1400	11	113	14.0

1 *Tensile strength x (y) means: "annealed (fully hardened)".

2 Nickel-alloys used over large temperature range. Properties vary with temperature.

2.4. POLYMERS

Polymers – Mechanical

Material	Density [$kg\ m^{-3}$]	Elastic Modulus [GPa]	Ultimate Tensile Strength[1] [MPa]	Application Notes
Thermoplastics				
Acrylics	1170-1200	2.7-3.5	48-76	Perspex, light fittings, lenses, signs, sanitaryware
Cellulose Acetate	1230-1340	0.4-2.7	13-59	Toys, electrical parts, packaging, cosmetics
Cellulose Acetate Butyrate	1150-1220	0.34-1.4	18-48	Panel lights, street signs, piping, reflectors
Polytetra fluoroethylene (PTFE)	2140-2200	0.35-0.62	17-42	Bearings, non-stick coatings, including pans, tubes and vessels for aggregate chemicals
Fluorinated Ethylene Propylene	2140-2170	0.35-0.48	17-24	Wire insulation, electronic components
Nylon 6	1110-1140	1.1-2.5	70-84	Moulded parts, low friction bearings/bushings, gears, wheels, cams, latches, fuel lines, electrical insulators
Nylon 66	1140-1150	1.8-2.8	48-84	
Polyacetal	1410-1420	2.8	49	Bearings, pipe fittings, gears, water pumps, load-bearing parts (Stiff, strong at relatively high temperatures)
Polycarbonate	1200	2.2	59-65	Machine housings, helmets, lamp covers, electrical equipment, safety glasses
Low Density Polyethylene (LDPE)	913-970	0.12-0.24	7-16	Bags, film packaging, electrical insulation
High Density Polyethylene (HDPE)	935-970	0.5-1.04	22-38	Sheet, tube, films and coatings, large moulded parts.

Polymers – Mechanical

Material	Density [kg m^{-3}]	Elastic Modulus [GPa]	Ultimate Tensile Strength[1] [MPa]	Application Notes
Polypropylene	900-910	0.9-1.38	29-38	Containers for food/products. Domestic appliances, furniture, toys, car parts, bristles
Polystyrene	1040-1050	2.8-4.14	35-84	Toys, electrical insulation, packaging, refrigerator insulation, CD cases
Acrylonitrile Butadiene Styrene (ABS)	1020-1090	0.7-2.8	17-62	Phone casings, hairdryers, TVs
Polyvinyl-Chloride (PVC)	1300-1580	2.4-4.1	59	Rainwater pipes, guards, ducts, window and door frames
Polysulphone	1240	70	70	Aircraft parts, circuit boards (strong, stiff, low creep)
Thermosets				
Epoxy – Cast	1150	1.4-4.1	35-84	Castings, electrical parts, protective coatings, circuit boards, adhesives
Epoxy – 60% Glass Fibres	1800	21-25	200-420	Boat structures, tabletops (strong composite material)
Melamine Formaldehyde	1500-1600	7-10	55-85	Handles, fittings, crockery
Phenol Formaldehyde	1240-1300	5-7	35-55	Circuit boards, gears, cams, brake lining matrix (filled with glass or metal powder)
Polyester (Unfilled)	1300	2.4	55	Boats (often as a composite with woven glass fibre)
Urea Formaldehyde (Cellulose-Filled)	1500-1600	7-13	50-80	Similar to melamine

1 Properties are temperature dependent. Properties at *Standard Conditions* used where possible (20°C, 101325 Pa).

Polymers – Thermal and Electrical

Material	Density [kg m^{-3}]	Melting Point [°C]	Coefficient of Thermal Expansion [10^{-6} K^{-1}]	Volume Resistivity [10^{9} Ω m^{-1}]
Thermoplastics				
Acrylics	1170-1200	2.7-3.5	50-90	$>10^{5}$
Cellulose Acetate	1230-1340	0.4-2.7	80-160	$10-10^{4}$
Cellulose Acetate Butyrate	1150-1220	0.34-1.4	110-170	$10-10^{3}$
Polytetrafluoroethylene (PTFE)	2140-2200	0.35-0.62	90-220	10^{10}
Fluorinated Ethylene Propylene	2140-2170	0.35-0.48	83-105	10^{10}
Nylon 6	1110-1140	1.1-2.5	80-130	$10^{3}-10^{6}$
Nylon 66	1140-1150	1.8-2.8	100-150	$4x10^{5}$
Polyacetal	1410-1420	2.8	181	$6x10^{5}$
Polycarbonate	1200	2.2	70	$2.1x10^{7}$
Low Density Polyethylene (LDPE)	913-970	0.12-0.24	160-180	$>10^{7}$
High Density Polyethylene (HDPE)	935-970	0.5-1.04	110-130	$>10^{7}$
Polypropylene	900-910	0.9-1.38	110	$>10^{7}$
Polystyrene	1040-1050	2.8-4.14	50-83	$>10^{4}$
Acrylonitrile Butadiene Styrene (ABS)	1020-1090	0.7-2.8	80-100	$10^{4}-10^{7}$
Polyvinyl-Chloride (PVC)	1300-1580	2.4-4.1	50-100	$10^{7}-10^{9}$
Polysulphone	1240	2.5	-	$10^{5}-10^{8}$
Thermosets				
Epoxy – Cast	1150	1.4-4.1	60	-
Epoxy – 60% Glass Fibres	1800	21-25	10-50	-
Melamine Formaldehyde	1500-1600	7-10	40	$10^{3}-10^{5}$
Phenol Formaldehyde	1240-1300	5-7	68	$10^{2}-10^{3}$
Polyester (Unfilled)	1300	2.4	55-100	-
Urea Formaldehyde (Cellulose-Filled)	1500-1600	7-13	-	-

1 Properties are temperature dependent. Properties at *Standard Conditions* used where possible (20°C, 101325 Pa).

2.5. SPECIFIC HEAT CAPACITY

Material	c_p $[kJ\,kg^{-1}K^{-1}]$	Material	c_p $[kJ\,kg^{-1}K^{-1}]$
Aluminium	0.897–0.913	Lead	0.128
Alumina Al_2O_3	0.451–0.955	Magnesium alloy	1.010
Air (dry, sea level)	1.005	Mercury	0.140
Alcohol, ethyl	2.440	Nickel	0.461–0.471
Ammonia, liquid	4.700	Nitrogen	1.040
Ammonia, gas	2.060	Oxygen	0.918
Asphalt	0.920	Paper	1.336
Bismuth	0.123	Polyethylene terephthalate	1.250
Brass	0.375	Polyisoprene natural rubber	1.880
Brick	0.840	Polyisoprene hard rubber	1.380
Bronze	0.370	Polymethylmethacrylate	1.500
Concrete	0.880	Polypropylene	1.920
Copper	0.386	Polystyrene	1.30–1.50
Brass	0.380	Polyurethane elastomer	1.800
Glass (crown)	0.670	Polyvinylchloride	0.84–1.17
Glass (pyrex)	0. 753	Silicon	0.705
Glass wool	0.840	Silicon carbide	0.670–0.678
Gold	0.129	Silver	0.235
Granite	0.790	Steel	0.490
Graphite carbon	0.717	Tin	0.228
Helium	5.193	Titanium	0.523
Hydrogen	14.304	Tungsten	0.134
Ice (-5 °C)	2.09	Water	4.186
Ice (-10 °C)	2.05	Wood	1.30–2.40
Iron	0.449	Zinc	0.387

A list of heat capacities for elementally-pure metallic solids can be found in the previous section on The Elements.

2.6. ASHBY CHARTS

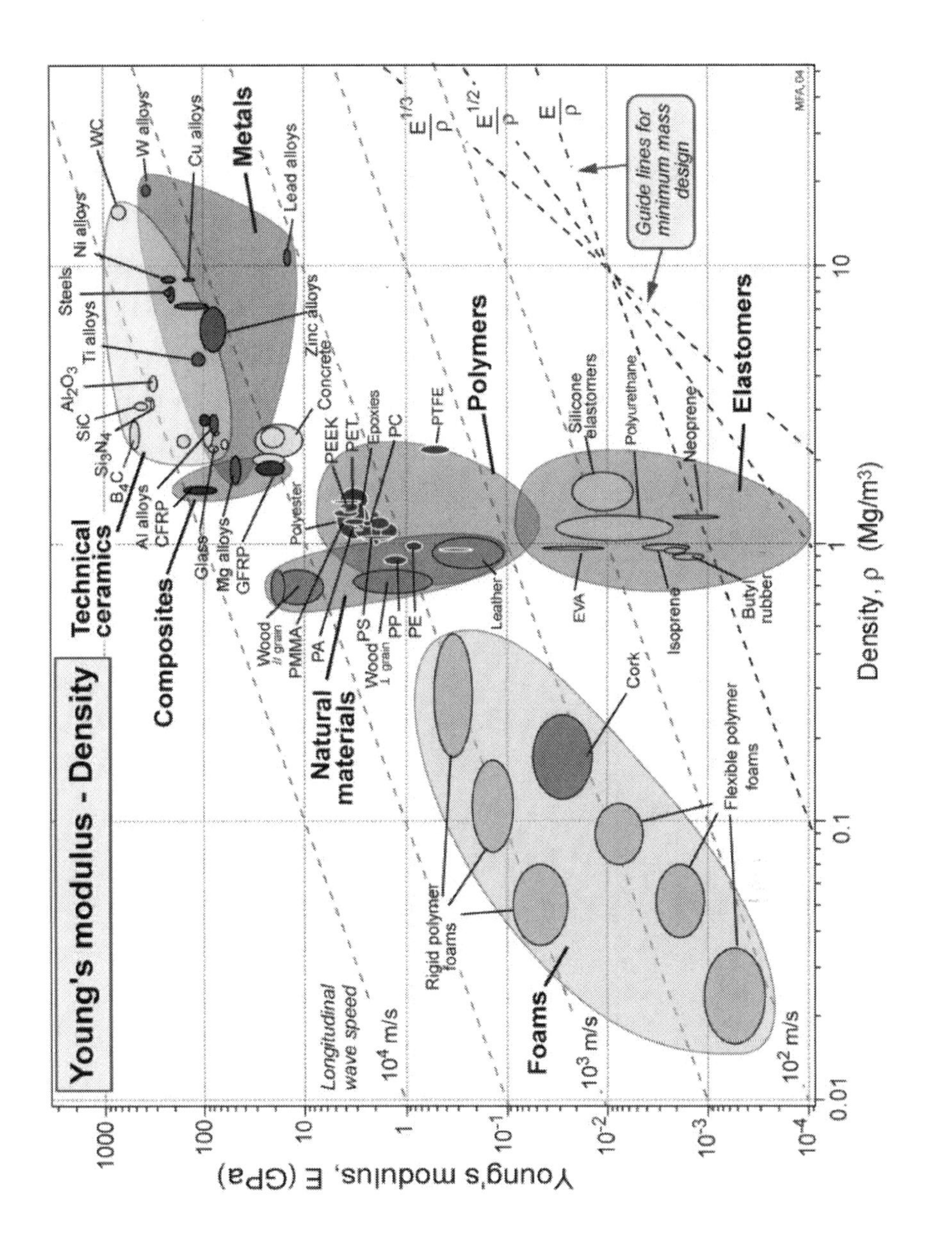

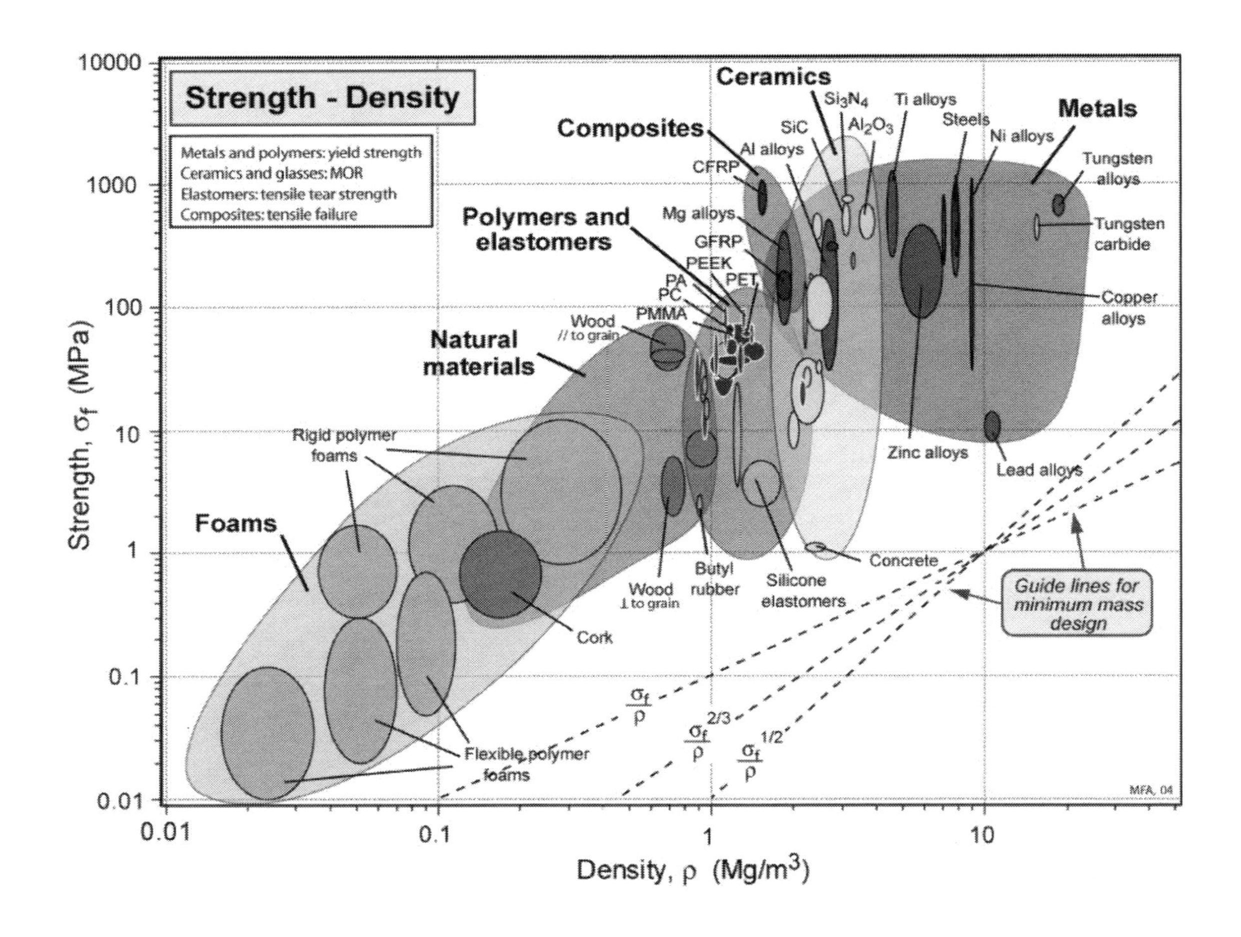
Strength - Density
Metals and polymers: yield strength
Ceramics and glasses: MOR
Elastomers: tensile tear strength
Composites: tensile failure
Metals
Ceramics
Composites
Polymers and elastomers
Natural materials
Foams
Tungsten alloys
Tungsten carbide
Copper alloys
Ni alloys
Steels
Ti alloys
Lead alloys
Zinc alloys
Concrete
Si_3N_4
Al_2O_3
SiC
Al alloys
CFRP
Mg alloys
GFRP
PEEK
PET
PA
PC
PMMA
Silicone elastomers
Butyl rubber
Wood ⊥ to grain
Wood // to grain
Cork
Rigid polymer foams
Flexible polymer foams
Guide lines for minimum mass design
$\frac{\sigma_f}{\rho}$
$\frac{\sigma_f^{2/3}}{\rho}$
$\frac{\sigma_f^{1/2}}{\rho}$
MFA, 04
Strength, σ_f (MPa)
Density, ρ (Mg/m^3)
10000
1000
100
10
1
0.1
0.01
0.01
0.1
1
10

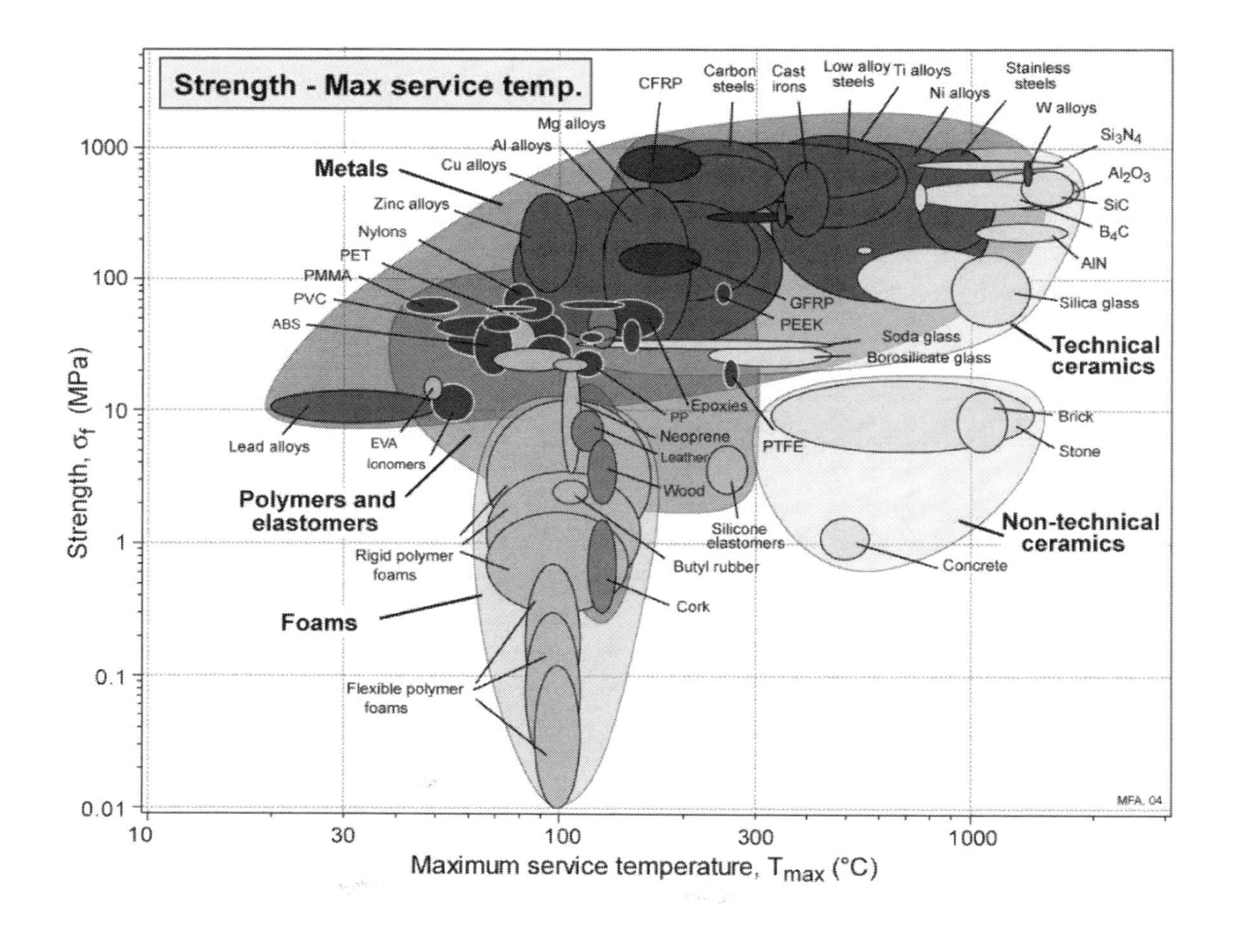
Strength - Max service temp.
CFRP
Carbon steels
Cast irons
Low alloy steels
Ti alloys
Ni alloys
Stainless steels
W alloys
Si_3N_4
Al_2O_3
SiC
B_4C
AlN
Silica glass
Technical ceramics
Mg alloys
Al alloys
Cu alloys
Metals
Zinc alloys
Nylons
PET
PMMA
PVC
ABS
GFRP
PEEK
Soda glass
Borosilicate glass
Lead alloys
EVA
Ionomers
PP
Epoxies
Neoprene
Leather
Wood
PTFE
Brick
Stone
Polymers and elastomers
Silicone elastomers
Non-technical ceramics
Rigid polymer foams
Butyl rubber
Concrete
Cork
Foams
Flexible polymer foams
MFA, 04
Strength, σ_f (MPa)
1000
100
10
1
0.1
0.01
10
30
100
300
1000
Maximum service temperature, T_{max} (°C)

2.7. COMPOSITES

Classifications

By matrix material	By dispersed (reinforcing) phase
Metal Matric	Particle
Ceramic Matrix	Fibre
Polymer Matrix	Laminate

Examples	Matrix	Reinforcement
Alumina-silica fibre	MMC	Ceramic particle
Aluminium - silicon carbide	MMC	Ceramic particle
Aluminum matrix - Boron fibre	MMC	Metalloid fibre
Boron fibre reinforced polymer	PMC	Metalloid fibre
Copper matrix – Tungsten carbide	MMC	Ceramic particle
Nickel matrix – Tungsten carbide	MMC	Ceramic particle
Concrete	CMC	Ceramic particle
SiC reinforced Aluminium	MMC	Ceramic fibre
SiC reinforced glass ceramic matrix	CMC	Ceramic fibre
SiC/SiC	CMC	Ceramic fibre
Glass reinforced polymer (GRP)	PMC	Ceramic fibre
Carbon fibre reinforced polymer (CFRP)	PMC	Ceramic fibre
Aramid polymer composites	PMC	Polymer fibre
Plywood	-	Laminate
Sandwich panels	-	Laminate with honeycomb core

MMC: Metal Matrix. CMC: Ceramic Matrix. PMC: Polymer Matrix.

Two-Phase Composites (e.g. concrete)

Modulus of Elasticity

$$E_u = E_m V_m + E_p V_p$$ UPPER BOUND

$$E_l = \frac{E_m E_p}{V_m E_p + V_p E_m}$$ LOWER BOUND

Subscripts

$m = matrix$
$p = particulate$

For a two-phase composite, Modulus of Elasticity lies between upper and lower limits given by E_u and E_l. The property changes as a function of particle volume percent.

Fibre-Reinforced Composites

$$m_c = \rho_c v_c = \rho_f v_f + \rho_m v_m$$ MASS

$$V_f = v_f / v_c, \quad V_m = v_m / v_c$$ VOLUME FRACTION

$$r_f = A_f / A_c, \quad r_m = A_m / A_c$$ AREA FRACTION

Subscripts

$c = composite\ (total)$
$f = reinforcing\ fibre$

Volume fraction = area fraction if all fibres are the same length.

$$l_C = \frac{\sigma_f d}{2\tau_c}$$ CRITICAL FIBRE LENGTH

$\sigma_f = fibre\ tensile\ strength$
$d = fibre\ diameter$
$\tau_c = shear\ strength\ of\ fibre\ matrix\ interface$

Longitudinal loading

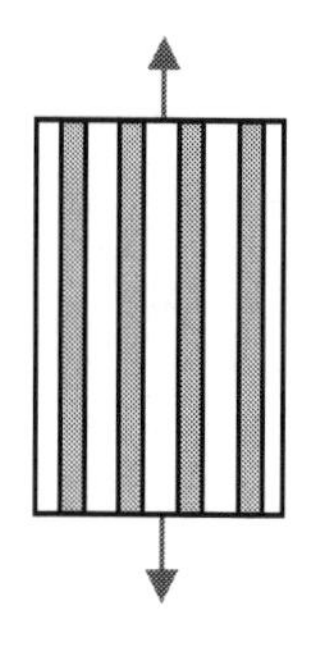

$E_l = E_f V_f + E_m V_m$ RULE OF MIXTURES

$F = \sigma A = \sigma_f A_f + \sigma_m A_m$

$\sigma = \sigma_f r_f + \sigma_m r_m$ LOAD CARRIED

$E_l = Young's\ modulus\ (longitudinal\ loading)$
$F = Force\ applied\ axially\ over\ section\ A\ [N]$
$\sigma = average\ cross\ section\ tensile\ stress\ [Pa]$
$A = total\ cross\ section\ area\ [m^2]$
$A_f = fibre\ cross\ section\ area\ [m^2]$

Transverse loading

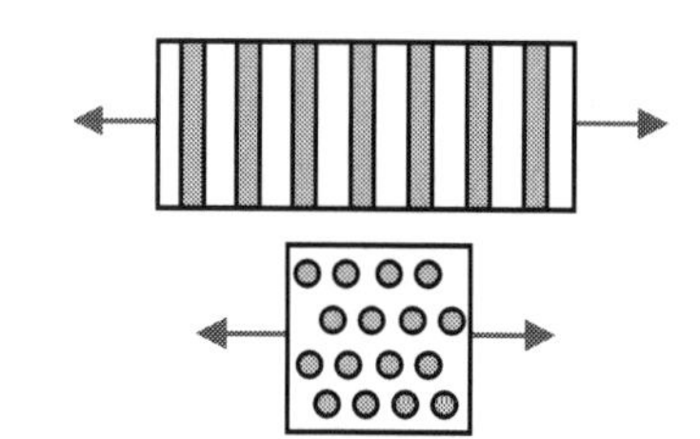

$$E_{ct} = \frac{E_m E_f}{V_m E_f + V_f E_m}$$

$E_{ct} = transverse\ modulus\ of\ elasticity$

For large particle composites, use the formula for transverse loading, replacing subscripts f with subscripts p to represent *Modulus of Elasticity*, and *Volume Fraction* of particulates.

	Longitudinal Tensile Strength (MPa)	**Transverse Tensile Strength (MPa)**
Glass-Polyester	700	20
Carbon (High Modulus)-Epoxy	1,000	35
Kevlar-Epoxy	1,200	20

Fibre content 50% approx. by vol.

Transverse loading of fibre composites in structural applications is highly undesirable.

Short Fibres, Random Orientation

$$E_{cd} = KE_f V_f + E_m V_m$$

$E_{cd} = Modulus\ of\ Elasticity\ for\ composite\ with\ short, randomly\ oriented\ fibres\ [Pa]$

Fibre orientation	Stress direction	Reinforcement Efficiency *K*
All fibres parallel	Parallel to fibres	1
	Perpendicular to fibres	0
Fibres randomly and uniformly distributed within a specified plane	Any direction in the plane of the fibres	$\frac{3}{8}$
Fibres randomly and uniformly distributed within three dimensions in space	Any direction	$\frac{1}{5}$

Discontinuous and Aligned Fibre Composites

When $l > l_c$ (long fibres)

$$\sigma_{cd}^* = \sigma_f^* V_f \left(1 - \frac{l_c}{2l}\right) + \sigma_m^{'}(1 - V_f)$$

When $l < l_c$ (short fibres)

$$\sigma_{cd}^* = \frac{l\tau_c}{d} V_f + \sigma_m^{'}(1 - V_f)$$

$\sigma_{cd}^* = longitudinal\ strength\ with\ discontinous\ and\ randomly\ oriented\ fibres\ [Pa]$
$\sigma_f^* = fibre\ fracture\ strength\ [Pa]$
$V_f = volume\ fraction\ of\ reinforcing\ fibre$
$l_c = critical\ fibre\ length\ [m]$
$\sigma_m^{'} = matrix\ stress\ at\ fibre\ failure\ [Pa]$
$\tau_c = the\ lower\ of\ fibre\ matrix\ bond\ strength\ or\ matrix\ shear\ strength\ [Pa]$

3. MECHANICS

3.1. STRESS-STRAIN RELATIONSHIP

Linear-elastic region

$$\sigma = E\varepsilon$$ STRESS

$E = Young's\ Modulus\ [Pa]$
$\sigma = stress\ [Pa]$
$\varepsilon = strain\ [dimensionless]$

Tensile or compressive stress σ is linearly proportional to its fractional extension or strain ε by the modulus of elasticity E. A rod of any elastic material obeys ***Hooke's Law*** (it behaves as a linear spring). The rod has length L and cross-sectional area A.

$$\sigma = \frac{F}{A}$$

$F = normal\ force\ [N]$
$A = cross\ section\ area\ [m^2]$

Saint-Venant's Principle: The difference between the effects of two different but statically equivalent loads becomes very small at sufficiently large distances from load. A uniform rod loaded in tension will have relatively uniform stress distribution throughout a cross section, except in the vicinity of its loaded ends.

$$\varepsilon = \frac{\Delta L}{L_0} = \frac{L - L_0}{L_0}$$ STRAIN

$L = length\ [m]$
$L_0 = original\ length\ [m]$

$$\nu = -\frac{lateral\ strain}{axial\ strain}$$ POISSON'S RATIO

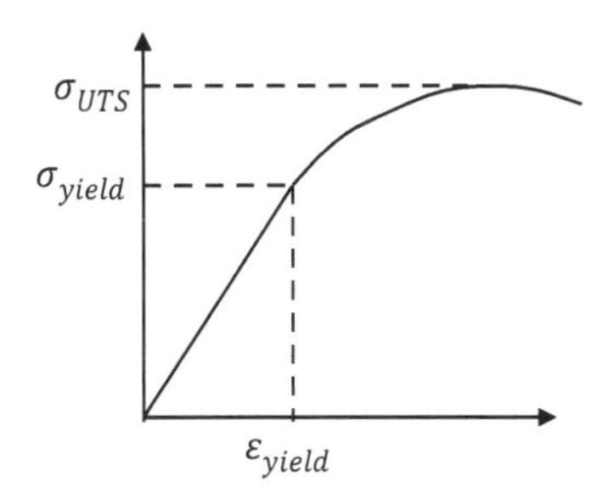

Uniaxial Stress

Isotropic materials

$$\varepsilon_1 = \frac{\sigma_1}{E}, \qquad \varepsilon_2 = \varepsilon_3 = -\nu\varepsilon_1$$

$\varepsilon_1 = strain\ along\ load\ axis$
$\varepsilon_2\ and\ \varepsilon_3 = strain\ along\ the\ other\ two\ mutually\ perpendicular\ axes$
$\nu = Poisson's\ ratio$

Shear

$$G = \frac{E}{2(1+\nu)}$$

SHEAR MODULUS

$E = Young's\ Modulus\ [GPa]$

The ***shear modulus*** (modulus of rigidity) is the ratio of shear stress to shear strain.

Linear-elastic region

$$\tau = G\gamma$$

$\tau = shear\ stress\ [Pa]$
$G = shear\ modulus\ [Pa]$
$\gamma = shear\ strain$

$$\tau = \frac{F_s}{A}$$

SHEAR STRESS

$F_s = shear\ force$
$A = cross\ section\ area\ (parallel\ to\ force)$

$$\gamma = \frac{\delta_x}{L}$$

SHEAR STRAIN

Triaxial Stress

Isotropic materials

$$K = \frac{E}{3(1-2\nu)}$$ BULK MODULUS

$K = Bulk\ Modulus\ [GPa]$
$E = Young's\ Modulus\ [GPa]$
$\nu = Poisson's\ Ratio$

> The Bulk Modulus, *K*, is the relative change in the volume of a body produced by a unit compressive or tensile stress acting uniformly over its surface. In most cases it is a measure of how resistant to compressibility that substance is.

Bulk Modulus for Common Materials

Material	$K\ [GPa]$	Material	$K\ [GPa]$
Air (adiabatic)	142	Steel	150
Air (constant pressure)	101	Petrol	1.07-1.49
Ethyl Alcohol	1.06	SAE 30 Oil	1.5
Glass	35-55	Seawater	2.34
Kerosine	1.3	Water	2.15
Methanol	0.823	Water Glycol	3.4

Strain

$$\varepsilon_x = \frac{1}{E}[\sigma_x - \nu(\sigma_y - \sigma_z)]$$

$$\varepsilon_y = \frac{1}{E}[\sigma_y - \nu(\sigma_z - \sigma_x)]$$

$$\varepsilon_z = \frac{1}{E}[\sigma_z - \nu(\sigma_x - \sigma_y)]$$

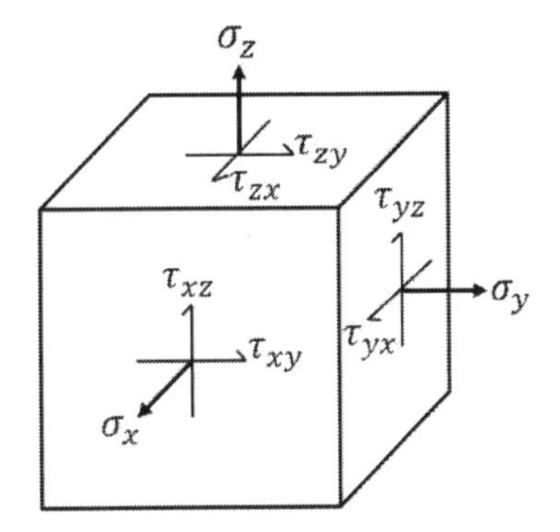

Shear Strain

$$\gamma_{xy} = \frac{\tau_{xy}}{G}$$

$$\gamma_{yz} = \frac{\tau_{yz}}{G}$$

$$\gamma_{zx} = \frac{\tau_{zx}}{G}$$

Cauchy Stress Tensor

$$[\sigma] = \begin{bmatrix} \sigma_{11} & \sigma_{12} & \sigma_{13} \\ \sigma_{21} & \sigma_{22} & \sigma_{23} \\ \sigma_{31} & \sigma_{32} & \sigma_{33} \end{bmatrix} = \begin{bmatrix} \sigma_x & \tau_{xy} & \tau_{xz} \\ \tau_{yx} & \sigma_y & \tau_{yz} \\ \tau_{zx} & \tau_{zy} & \sigma_z \end{bmatrix}$$

The **Cauchy stress tensor** is a second order tensor with nine components that completely define the state of stress at a point inside a material in the deformed state.

Stress Vector

$$\vec{T}^{(n)} = \hat{n} \cdot [\sigma]$$

$$T_j^{(n)} = \sigma_{ij} n_i$$

The **eigenvectors** of the stress tensor are the principal directions. The associated **eigenvalues** are the principal stresses (which are invariant; they do not depend on the arbitrary orientation chosen for the infinitesimal stress element).

$\vec{T}^{(n)} = stress\ vector\ (crossing\ an\ imaginary\ surface\ of\ an\ infinitesimal\ stress\ element)$
$n = unit\ vector\ normal\ to\ surface$

Magnitude of Normal Stress

$$\sigma_n = \vec{T}^{(n)} \cdot \hat{n} = T_i^{(n)} n_i = \sigma_{ij} n_i n_j$$

Magnitude of Shear Stress

$$\tau_n = \sqrt{(\vec{T}^{(n)})^2 - \sigma_n^2} = \sqrt{T_i^{(n)} T_i^{(n)} - \sigma_n^2}$$

$$where\ (\vec{T}^{(n)})^2 = T_i^{(n)} T_i^{(n)} = (\sigma_{ij} n_j)(\sigma_{ik} n_k) = \sigma_{ij}\sigma_{ik} n_j n_k$$

Infinitesimal Strain Tensor

$$[\varepsilon] = \begin{bmatrix} \varepsilon_{xx} & \frac{1}{2}\gamma_{xy} & \frac{1}{2}\gamma_{xz} \\ \frac{1}{2}\gamma_{yx} & \varepsilon_{yy} & \frac{1}{2}\gamma_{yz} \\ \frac{1}{2}\gamma_{zx} & \frac{1}{2}\gamma_{zy} & \varepsilon_{zz} \end{bmatrix}$$

$$where\ \varepsilon_{xy} = \frac{1}{2}\gamma_{xy} = \frac{1}{2}\left(\frac{\partial u_x}{\partial y} + \frac{\partial u_y}{\partial x}\right)\ etc.$$

3.2. PRESSURE VESSELS

Cylindrical Vessels

The state of stress in a thin-walled vessel is two dimensional *(i.e.* ***plane stress****)*. All shear stresses are zero, and the radial stress σ_{rr} varies from p on the inner surface to zero on the outer surface. The wall is assumed to be very thin compared to the other dimensions of the vessel: $t/R \ll 1$, and the internal pressure is assumed to be higher than the external (or atmospheric) pressure, else wall buckling may be a failure mode.

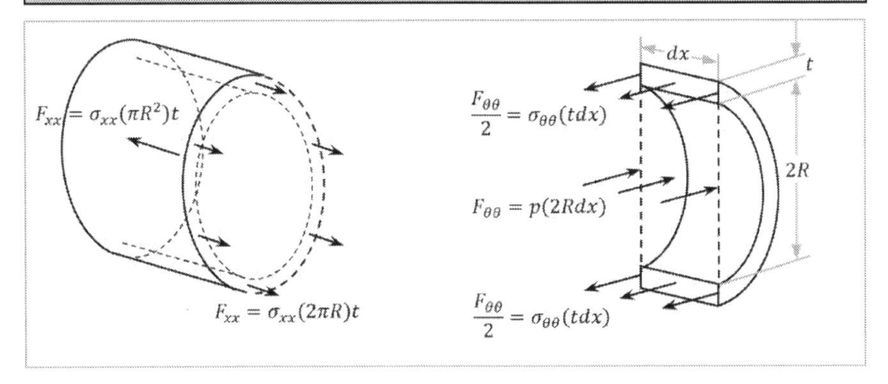

Stress Components

Hoop Stress

$$\sigma_{\theta\theta} = \frac{pR}{t}$$

Note that a hoop experiences the greatest stress at its inside, hence cracks in pipes may start from *inside* the pipe.

Axial Stress

$$\sigma_{xx} = \frac{pR}{2t}$$

$\sigma_{\theta\theta} = hoop\ stress\ [kPa]$
$\sigma_{xx} = axial\ stress\ [kPa]$

Stress Tensor

$$[\sigma] = \begin{bmatrix} \sigma_{xx} & \tau_{x\theta} & \tau_{xr} \\ \tau_{\theta x} & \sigma_{\theta\theta} & \tau_{\theta r} \\ \tau_{rx} & \tau_{r\theta} & \sigma_{rr} \end{bmatrix}$$

$$= \begin{bmatrix} \sigma_{xx} & 0 & 0 \\ 0 & \sigma_{\theta\theta} & 0 \\ 0 & 0 & 0 \end{bmatrix} = \frac{pR}{2t}\begin{bmatrix} 1 & 0 & 0 \\ 0 & 2 & 0 \\ 0 & 0 & 0 \end{bmatrix}$$

$p = gauge\ pressure\ [kPa]$
$R = inner\ radius\ [m]$
$t = wall\ thickness\ [m]$

A pressure vessel constructed of an isotropic material will often be twice as strong as it needs to be in the axial direction. Use reinforcing circumferential hoops or fibres with higher tensile strength for weight-saving.

If the cylinder and the end caps were to deform independently of each other, they would each tend to expand by different amounts. But since physical continuity of the wall must be maintained, the necessary adjustment in the displacement produces local bending stresses and shear stresses near to where the end cap joins the cylinder.

Spherical Vessels

A sphere is the optimal geometry for a closed pressure vessel. In the cylindrical vessel the internal pressure is resisted most effectively by the hoop stress. In the spherical vessel, the double curvature means that all stress directions around the stress element contribute to resisting the pressure; theoretically it can withstand twice the pressure.

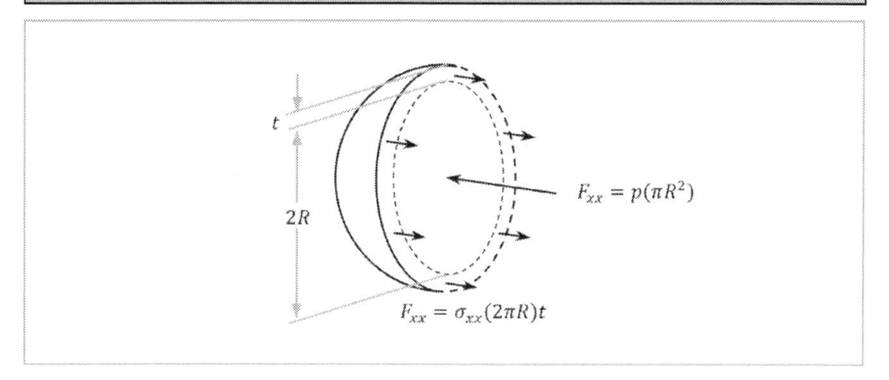

Stress Components

Wall Stress

$$\sigma = \frac{pR}{2t}$$

$\sigma = \sigma_{\theta\theta} = \sigma_{\phi\phi} = wall\ stress\ [kPa]$
$p = gauge\ pressure\ [kPa]$
$R = inner\ radius\ [m]$
$t = wall\ thickness\ [m]$

Stress Tensor

$$[\sigma] = \begin{bmatrix} \sigma_{\theta\theta} & \tau_{x\theta} & \tau_{\theta r} \\ \tau_{\phi\theta} & \sigma_{\phi\phi} & \tau_{\phi r} \\ \tau_{r\theta} & \tau_{r\phi} & \sigma_{rr} \end{bmatrix}$$

$$= \begin{bmatrix} \sigma & 0 & 0 \\ 0 & \sigma & 0 \\ 0 & 0 & 0 \end{bmatrix} = \frac{pR}{2t}\begin{bmatrix} 1 & 0 & 0 \\ 0 & 1 & 0 \\ 0 & 0 & 0 \end{bmatrix}$$

The stresses in any two orthogonal circumferential directions are the same, denoted σ.

Do not rely on thin-walled calculations for any significant engineering project, and especially not for safety critical applications. Consult ASME boiler and pressure vessel code (BPVC) before undertaking any practical design of pressure vessels.

3.3. MOHR'S CIRCLE

For Plane Stress

The normal to the stress-free surface is the z-direction.

$$\sigma_z = \tau_{zx} = \tau_{zy} = 0$$

$$[\sigma] = \begin{bmatrix} \sigma_x & \tau_{xy} \\ \tau_{yx} & \sigma_y \end{bmatrix}$$

Plane-Stress Transformation Equations

Normal Stresses on an oblique plane at any angle θ:

$$\sigma_n = \frac{\sigma_x + \sigma_y}{2} + \frac{\sigma_x - \sigma_y}{2}\cos 2\theta + \tau_{xy}\sin 2\theta$$

$$\tau_n = \frac{\sigma_x - \sigma_y}{2}\sin 2\theta + \tau_{xy}\cos 2\theta$$

Angle of the oblique plane containing principal stresses

$$\tan 2\theta_p = \frac{2\tau_{xy}}{\sigma_x - \sigma_y}$$

> An angle of θ on the stress element in physical space $=2\theta$ on the Mohr's Circle diagram.

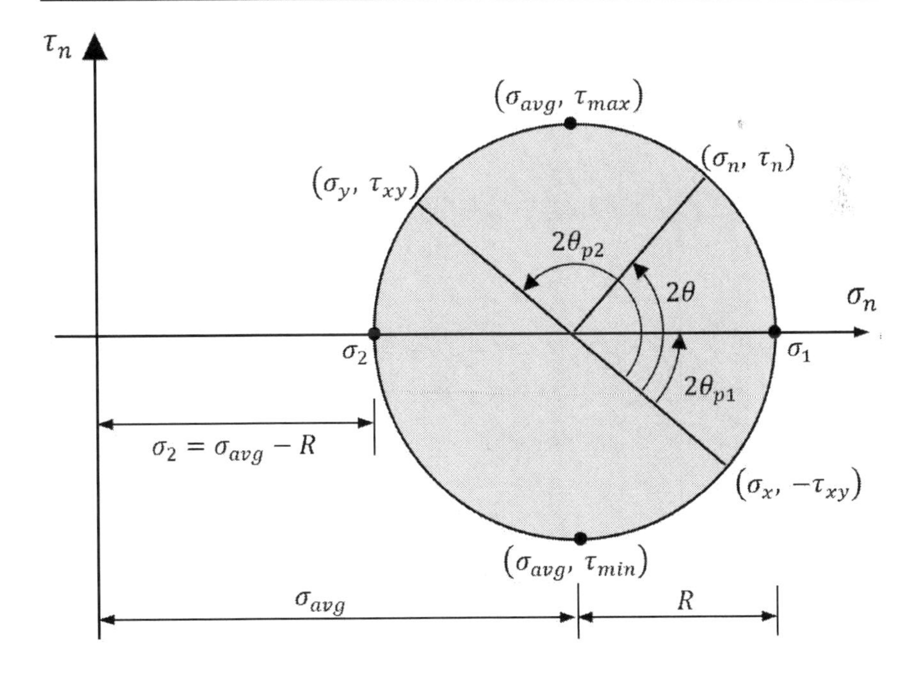

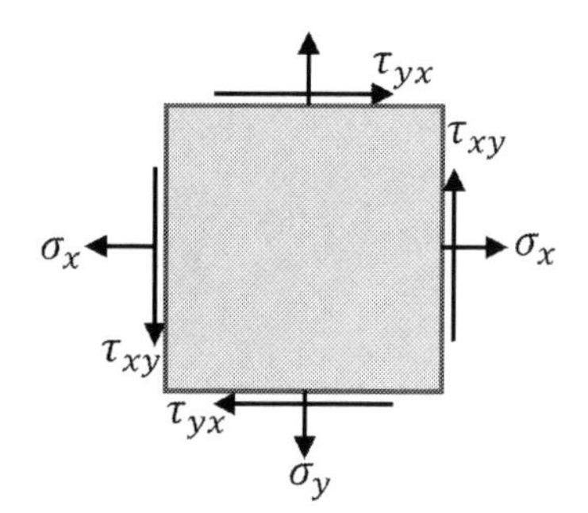

Infinitesimal element subject to arbitrary plane stress (normal and shear stress)

$$\tau_{yx} = \tau_{xy}$$

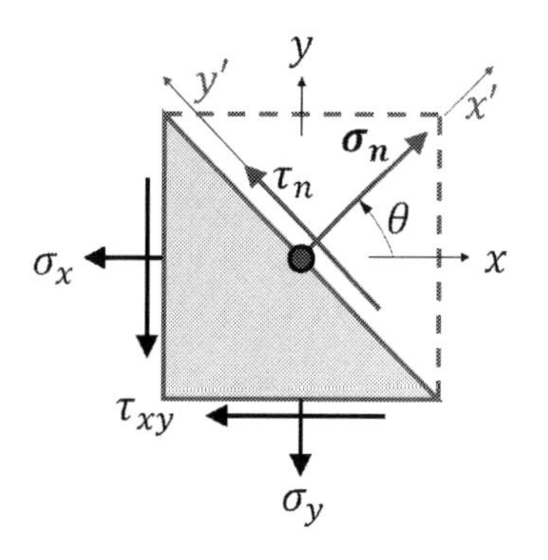

Stress components in physical space at an arbitrary plane passing through a point in a continuum under plane stress conditions.

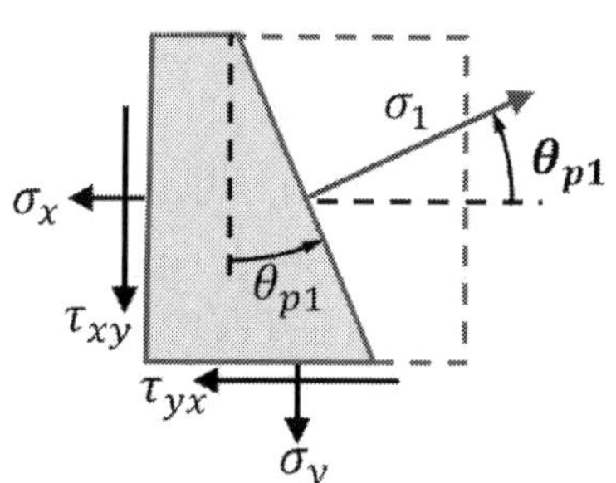

Plane containing principal stress σ_1 is inclined at angle σ_{p1} in physical space

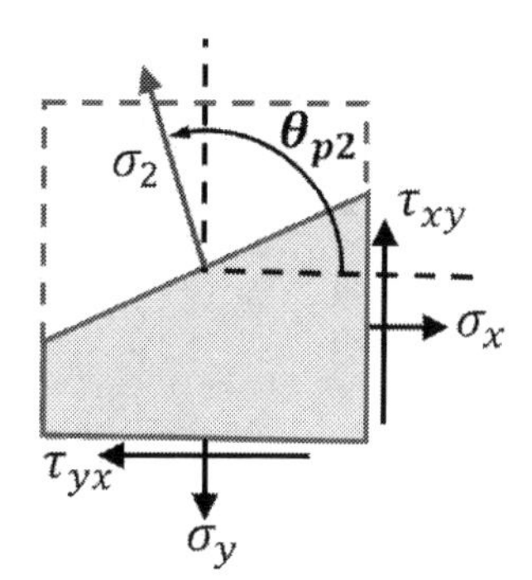

Plane containing principal stress σ_2 is inclined at angle σ_{p2} in physical space

Principal Stress

Given the state of stress in the $x-y$ axis, from the geometry of Mohr's Circle you can calculate principal normal stress and shear stress:

$$\sigma_1, \sigma_2 = \frac{\sigma_x + \sigma_y}{2} \pm \sqrt{\left(\frac{\sigma_x - \sigma_y}{2}\right)^2 + \tau_{xy}^2}$$

$$\tau_1, \tau_2 = \pm \sqrt{\left(\frac{\sigma_x - \sigma_y}{2}\right)^2 + \tau_{xy}^2}$$

The surfaces containing principal stresses σ_1, σ_2 have zero shear stresses, i.e. the plane containing principal stresses is under pure tension and compression only. The maximum shear stress occurs when $\sigma_1 = \sigma_2$.

Shear stress is maximum at 90° on Mohr's Circle, i.e. 45° on the material element. Ductile materials have a tendency to fail in shear along a plane at 45° to an external load.

Stress element on a rod in tension and torsion

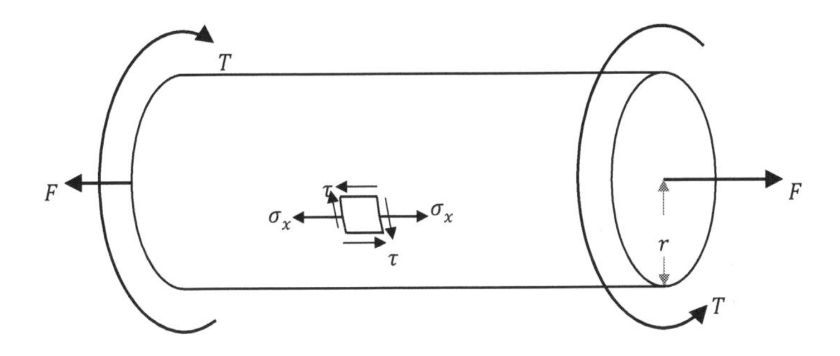

$$\sigma_x = \frac{F}{A} = \frac{F}{\pi r^2}$$

$$\tau = \frac{Tr}{J}, J = \frac{\pi d^4}{32} = \frac{\pi r^4}{2}$$

Failure occurs if any one of the stresses exceeds the allowable stress (e.g. σ_{yield}) for the material; This is the maximum stress criterion. For ductile materials, the effect of strain is not negligible, so also consider the von Mises yield criterion (maximum distortion energy criterion).

General Three-Dimensional Stress

Three Principal Shear Stresses

$$\tau_{max} = \tau_{1/3} = \text{ diameter of larger circle}$$

$$\tau_{1/2} = \frac{\sigma_1 - \sigma_2}{2},\ \tau_{2/3} = \frac{\sigma_2 - \sigma_3}{2},\ \tau_{1/3} = \frac{\sigma_1 - \sigma_3}{2}$$

Order of principal stresses $\sigma_1 > \sigma_2 > \sigma_3$.

$$\tau_{max} = \frac{1}{2}(\sigma_1 - \sigma_3)$$

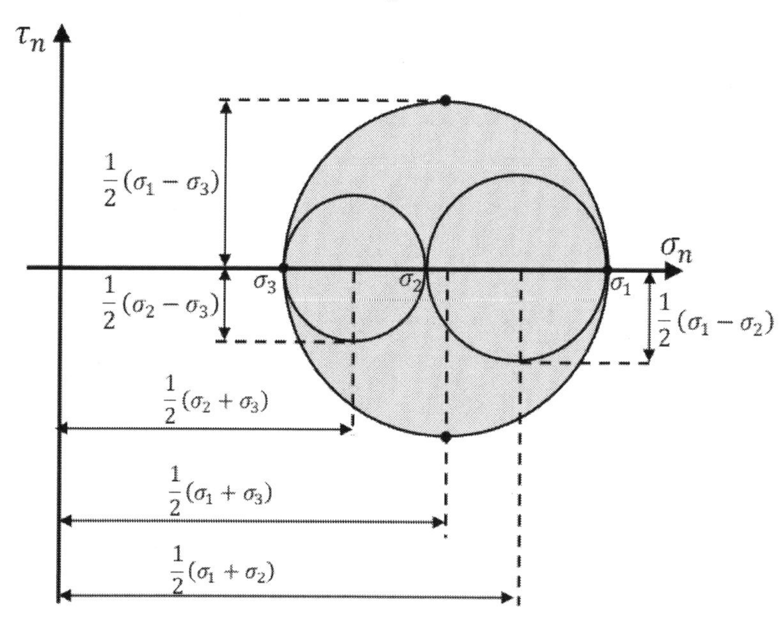

Yield Criteria

VON MISES

$$(\sigma_1 - \sigma_2)^2 + (\sigma_2 - \sigma_3)^2 + (\sigma_3 - \sigma_1)^2 = 2S_y^2$$

$S_y = \text{elastic limit of material in tensile test}$

Yield occurs if the principal stresses satisfy the von Mises' equation.

4. STRUCTURES

4.1. BENDING OF BEAMS

Second Moment of Area

$$I = \iint_A r^2 dA$$

$$I_{xx} = \iint_A y^2 \, dx \, dy$$

$I = second\ moment\ of\ area = area\ moment\ of\ inertia\ (about\ a\ specified\ axis) [m^4]$
$y = perpendicular\ distance\ from\ axis\ x\ to\ the\ element\ dA\ [m]$
$dA = inifintesimal\ area\ [m^2]$

Product Moment of Area

$$I_{xy} = \iint_A yx \, dx \, dy$$

Parallel Axis Theorem

$$I_{x'} = I_x + A\delta^2$$

$\delta = perpendicular\ distance\ between\ x\ and\ x'\ axes$

Radius of Gyration

$$R_g = \sqrt{\frac{I}{A}}$$

$$I = AR_g^{\ 2}$$

$R_g = radius\ of\ gyration\ [m]\ (occasionally\ written\ as\ k\ or\ r)$

The **radius of gyration** is used in estimating the stiffness of a column and describes the distribution of area at a cross-section (not to be confused with the mass radius of gyration in classical mechanics). If the principal moments of the two-dimensional gyration tensor are not equal, the column will tend to buckle around the axis with the smaller principal moment. For example, a column with an elliptical cross-section will tend to buckle in the direction of the smaller semi-axis.

4.2. THE BENDING FORMULA

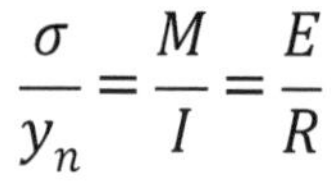

$$\frac{\sigma}{y_n} = \frac{M}{I} = \frac{E}{R}$$

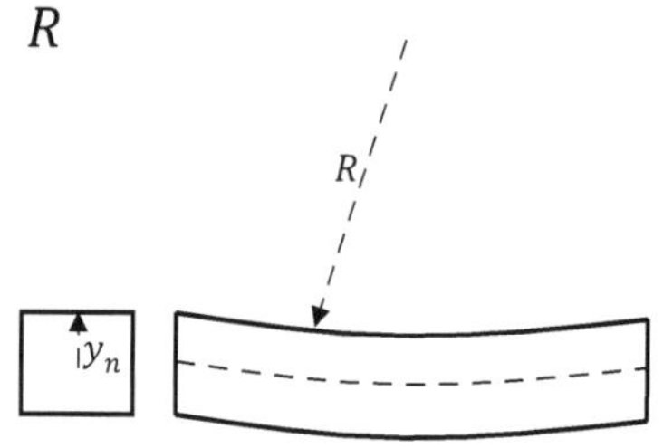

$\sigma = stress\ in\ the\ beam\ at\ position\ y_n$
$y_n = distance\ from\ neutral\ axis\ of\ beam$
$M = bending\ moment$
$I = second\ moment\ of\ area$
$E = Young's\ Modulus$
$R = radius\ of\ curvature$

For small slopes:

$$\theta = \frac{dy}{dx} = \frac{1}{EI}\int M dx$$ SLOPE

$$y = \frac{1}{EI}\iint M dx dx$$ DEFLECTION

$$\frac{1}{R} = -\frac{d^2y}{dx^2} = \frac{M}{EI}$$ CURVATURE

$$M = -EI\frac{d^2y}{dx^2}$$ BENDING MOMENT

$$V = -\frac{d}{dx}\left(EI\frac{d^2y}{dx^2}\right) = -\frac{dM}{dx}$$ SHEAR FORCE

$$w(x) = -\frac{d^2M}{dx^2} = \frac{dV}{dx}$$ LOADING

$\theta = slope\ [rad]$
$w = load\ per\ unit\ length\ [N\ m^{-1}]$
$V = shear\ force\ [N]$

Plane Sections

Section	Second moment of area $I_{xx} = Ak_{xx}^2$	Polar 2nd moment of area about z $J_{zz} = Ak_{zz}^2$	Elastic Section Modulus Z
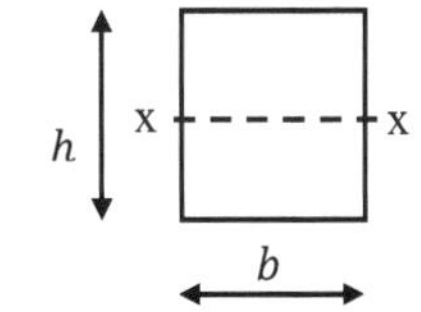 Rectangle (h, b, x–x)	$\frac{bh^3}{12}$	$\frac{bh(b^2+h^2)}{12}$	$\frac{bh^2}{6}$
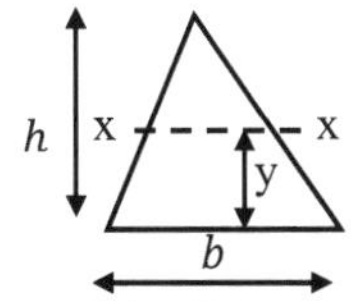 Triangle (h, b, y, x–x)	$\frac{bh^3}{36}$	-	-
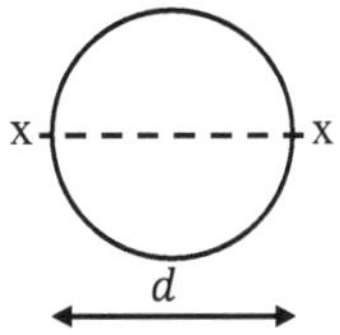 Circle (d, x–x)	$\frac{\pi d^4}{64} = \frac{\pi r^4}{4}$	$\frac{\pi d^4}{32} = \frac{\pi r^4}{2}$	$\frac{\pi r^3}{4} = \frac{\pi d^3}{32}$
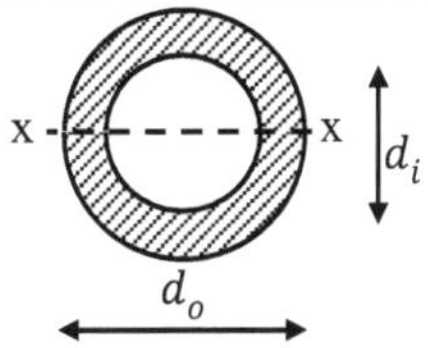 Hollow circle (d_o, d_i, x–x)	$\frac{\pi(d_o^4 - d_i^4)}{64} = \frac{\pi(r_o^4 - r_i^4)}{4}$	$\frac{\pi(d_o^4 - d_i^4)}{32} = \frac{\pi(r_o^4 - r_i^4)}{2}$	$\frac{\pi(d_o^4 - d_i^4)}{32d_o}$
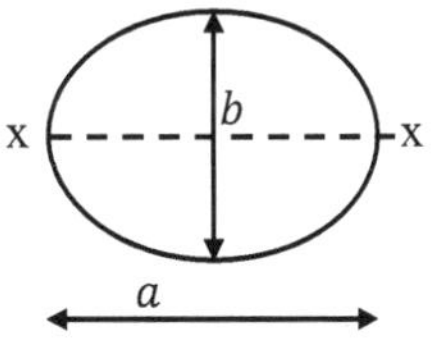 Ellipse (a, b, x–x)	$\frac{\pi ab^3}{4}$	$\frac{\pi ab(a^2+b^2)}{4}$	-
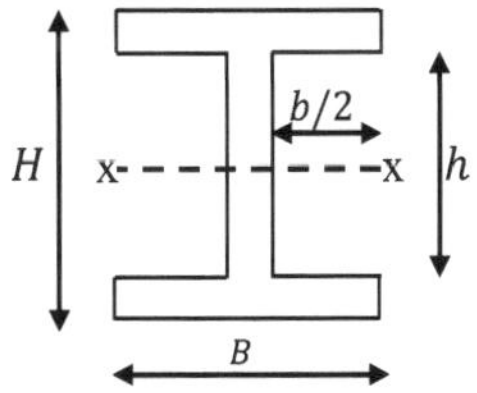 I-section (H, h, B, b/2, x–x)	$\frac{BH^3 - bh^3}{12}$	-	$\frac{BH^2}{6} - \frac{bh^3}{6H}$

4.3. LINEAR ELASTIC BEAMS

Cantilever End load

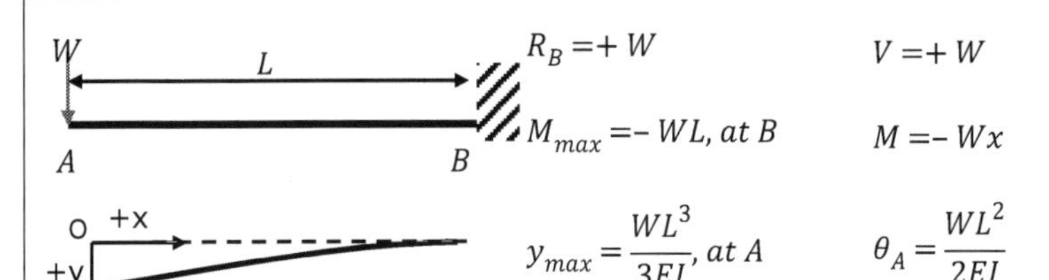

$R_B = + W$

$M_{max} = - WL, at\ B$

$y_{max} = \frac{WL^3}{3EI}, at\ A$

$V = + W$

$M = - Wx$

$\theta_A = \frac{WL^2}{2EI}$

Cantilever Concentrated load W, distance b from A

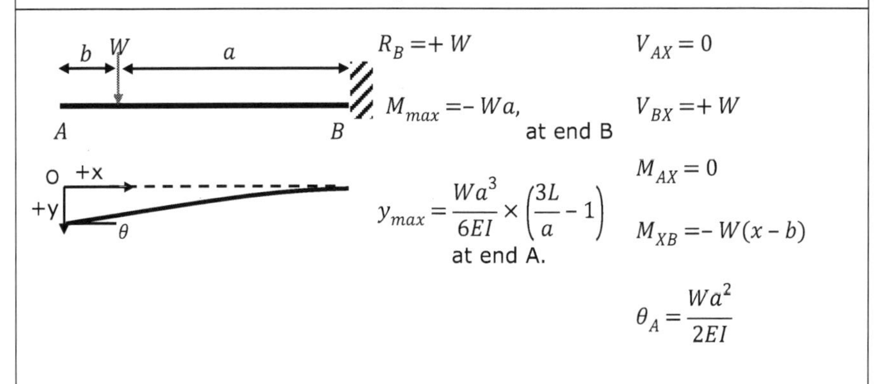

$R_B = + W$

$M_{max} = - Wa,$ at end B

$y_{max} = \frac{Wa^3}{6EI} \times \left(\frac{3L}{a} - 1\right)$ at end A.

$V_{AX} = 0$

$V_{BX} = + W$

$M_{AX} = 0$

$M_{XB} = - W(x - b)$

$\theta_A = \frac{Wa^2}{2EI}$

Cantilever Uniform load

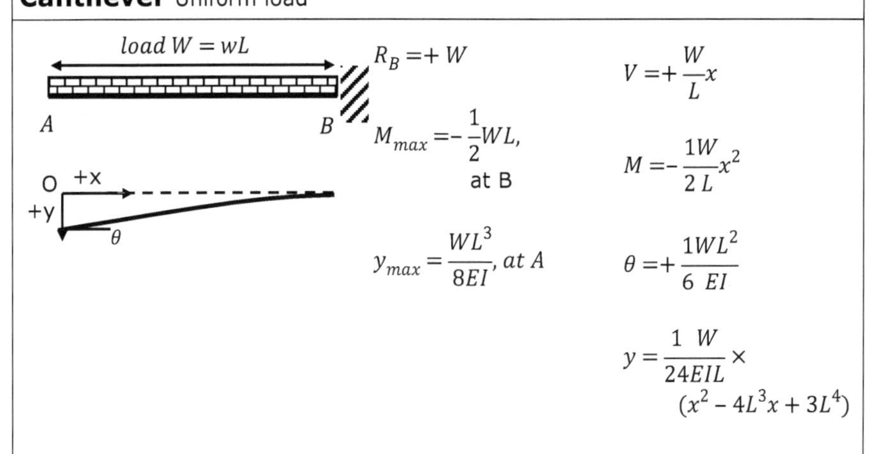

$R_B = + W$

$M_{max} = - \frac{1}{2}WL,$ at B

$y_{max} = \frac{WL^3}{8EI}, at\ A$

$V = + \frac{W}{L}x$

$M = - \frac{1W}{2\,L}x^2$

$\theta = + \frac{1WL^2}{6\ EI}$

$y = \frac{1\ W}{24EIL} \times (x^2 - 4L^3x + 3L^4)$

Simply supported beam Concentrated load W at centre

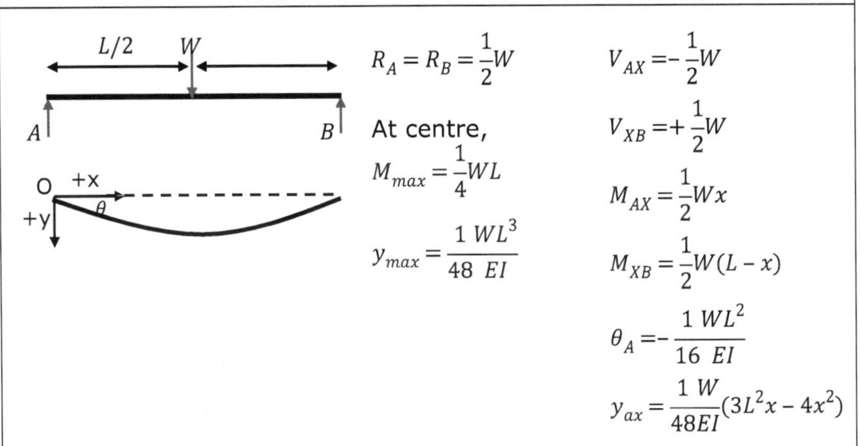

$$R_A = R_B = \frac{1}{2}W$$

At centre,

$$M_{max} = \frac{1}{4}WL$$

$$y_{max} = \frac{1}{48}\frac{WL^3}{EI}$$

$$V_{AX} = -\frac{1}{2}W$$

$$V_{XB} = +\frac{1}{2}W$$

$$M_{AX} = \frac{1}{2}Wx$$

$$M_{XB} = \frac{1}{2}W(L - x)$$

$$\theta_A = -\frac{1}{16}\frac{WL^2}{EI}$$

$$y_{ax} = \frac{1}{48EI}\frac{W}{}(3L^2x - 4x^2)$$

Simply supported beam Concentrated load W, distance a from x=0

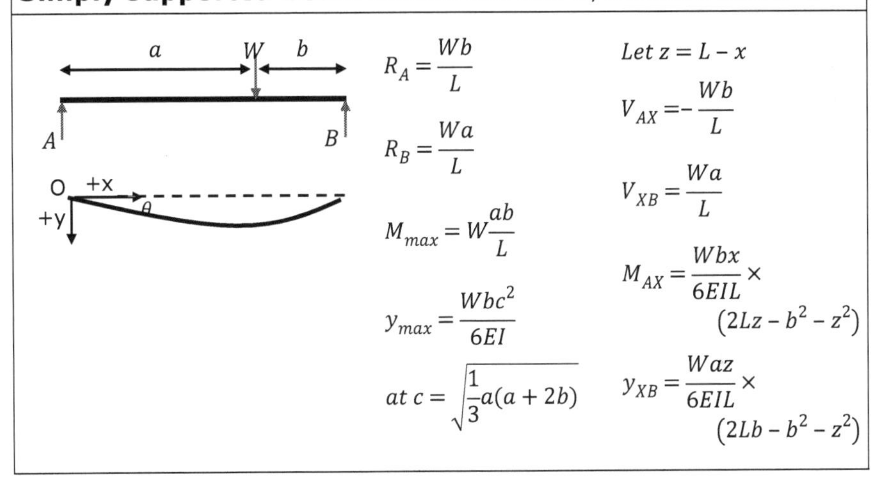

$$R_A = \frac{Wb}{L}$$

$$R_B = \frac{Wa}{L}$$

$$M_{max} = W\frac{ab}{L}$$

$$y_{max} = \frac{Wbc^2}{6EI}$$

$$at\ c = \sqrt{\frac{1}{3}a(a + 2b)}$$

$$Let\ z = L - x$$

$$V_{AX} = -\frac{Wb}{L}$$

$$V_{XB} = \frac{Wa}{L}$$

$$M_{AX} = \frac{Wbx}{6EIL} \times (2Lz - b^2 - z^2)$$

$$y_{XB} = \frac{Waz}{6EIL} \times (2Lb - b^2 - z^2)$$

Simply supported beam Uniform load

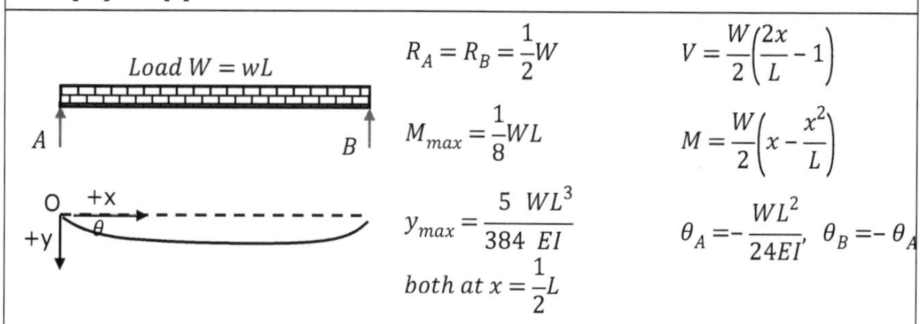

$$R_A = R_B = \frac{1}{2}W$$

$$M_{max} = \frac{1}{8}WL$$

$$y_{max} = \frac{5}{384}\frac{WL^3}{EI}$$

$$both\ at\ x = \frac{1}{2}L$$

$$V = \frac{W}{2}\left(\frac{2x}{L} - 1\right)$$

$$M = \frac{W}{2}\left(x - \frac{x^2}{L}\right)$$

$$\theta_A = -\frac{WL^2}{24EI},\quad \theta_B = -\theta_A$$

Simply supported beam Intermediate couple

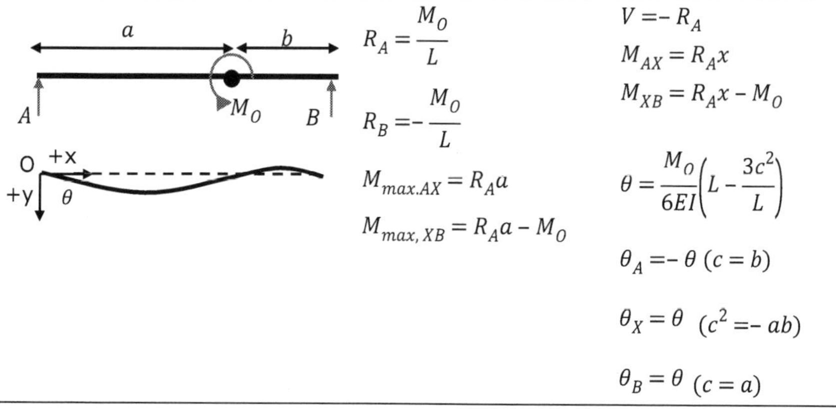

$$R_A = \frac{M_O}{L}$$

$$R_B = -\frac{M_O}{L}$$

$$M_{max.AX} = R_A a$$

$$M_{max,XB} = R_A a - M_O$$

$$V = -R_A$$

$$M_{AX} = R_A x$$

$$M_{XB} = R_A x - M_O$$

$$\theta = \frac{M_O}{6EI}\left(L - \frac{3c^2}{L}\right)$$

$$\theta_A = -\theta\ (c = b)$$

$$\theta_X = \theta\ \ (c^2 = -ab)$$

$$\theta_B = \theta\ (c = a)$$

Both ends fixed Central load

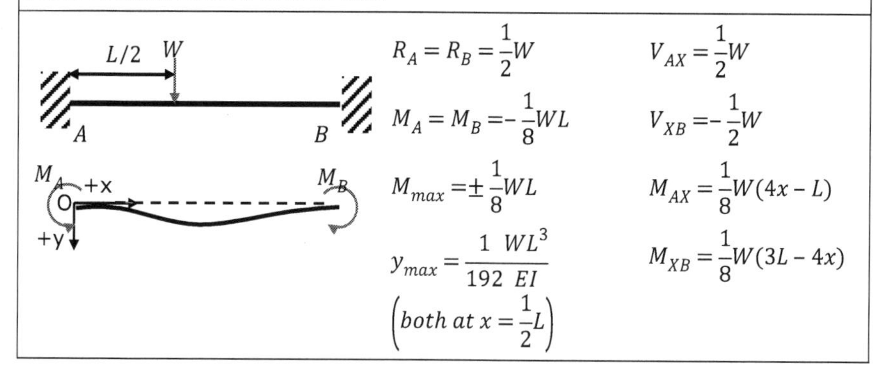

$$R_A = R_B = \frac{1}{2}W$$

$$M_A = M_B = -\frac{1}{8}WL$$

$$M_{max} = \pm\frac{1}{8}WL$$

$$y_{max} = \frac{1}{192}\frac{WL^3}{EI}$$

$$\left(both\ at\ x = \frac{1}{2}L\right)$$

$$V_{AX} = \frac{1}{2}W$$

$$V_{XB} = -\frac{1}{2}W$$

$$M_{AX} = \frac{1}{8}W(4x - L)$$

$$M_{XB} = \frac{1}{8}W(3L - 4x)$$

Both ends fixed Distributed load

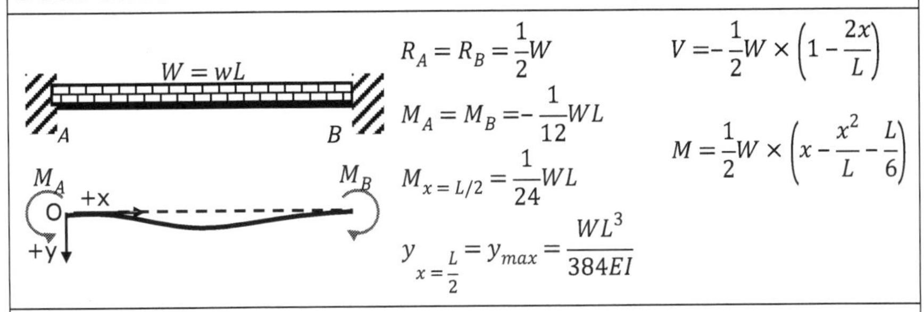

$$R_A = R_B = \frac{1}{2}W$$

$$M_A = M_B = -\frac{1}{12}WL$$

$$M_{x=L/2} = \frac{1}{24}WL$$

$$y_{x=\frac{L}{2}} = y_{max} = \frac{WL^3}{384EI}$$

$$V = -\frac{1}{2}W \times \left(1 - \frac{2x}{L}\right)$$

$$M = \frac{1}{2}W \times \left(x - \frac{x^2}{L} - \frac{L}{6}\right)$$

Effect of end rotation Pinned joints

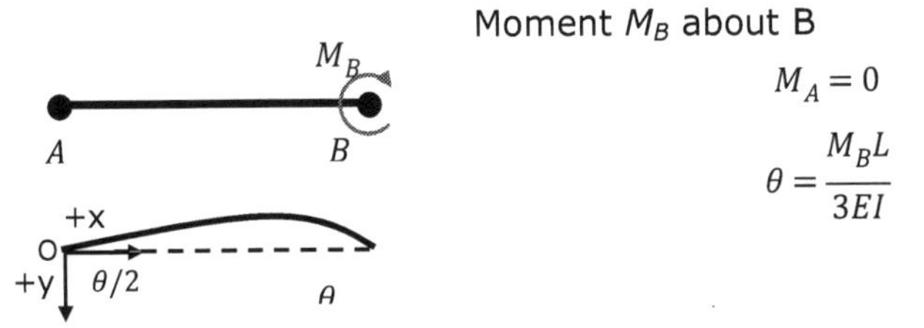

Moment M_B about B

$$M_A = 0$$

$$\theta = \frac{M_B L}{3EI}$$

Effect of end rotation Fixed-pinned

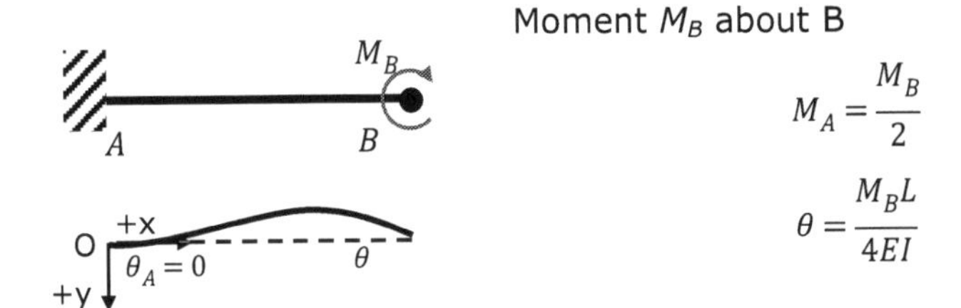

Moment M_B about B

$$M_A = \frac{M_B}{2}$$

$$\theta = \frac{M_B L}{4EI}$$

The equations for Linear Elastic Beams are valid for small deflections (neglecting *shear deformation)*, i.e. valid if $y < depth/2$.

$R_A = left\ reaction\ force\ [N]$
$R_B = right\ reaction\ force\ [N]$
$M = bending\ moment\ [N\ m]$
$W = concentrated\ load\ [N]$
$y = \max\ deflection\ [m]$
$\theta = End\ slope\ [radians]\ between\ beam\ and\ x - axis.$
$V = shear\ force\ [N\ m^{-2}]$

4.4. TORSION OF SHAFTS

$$\frac{\tau}{r} = \frac{T}{J} = \frac{G\theta}{L}$$

$\tau = shear\ stress\ [kPa]$
$r = radius\ from\ neutral\ axis\ [m]$
$T = torsion\ [N \cdot m]$
$J = polar\ second\ moment\ of\ area\ [m^4]$
$G = shear\ modulus\ [kPa]$
$\theta = angle\ of\ twist\ [rad]$
$L = length\ of\ shaft\ [m]$

Max Shear Stress

$$\tau_{max} = \frac{Tr}{J} = \frac{Td}{2J}$$

$r = radius\ of\ shaft\ [m]$
$d = diameter\ of\ shaft\ [m]$

For a hollow shaft:

$$J = \frac{\pi\left(d_o^4 - d_i^4\right)}{32}$$

For a round shaft:

$$J = \frac{\pi d^4}{32}$$

$$\tau_{max} = \frac{2T}{\pi r_o^3}$$

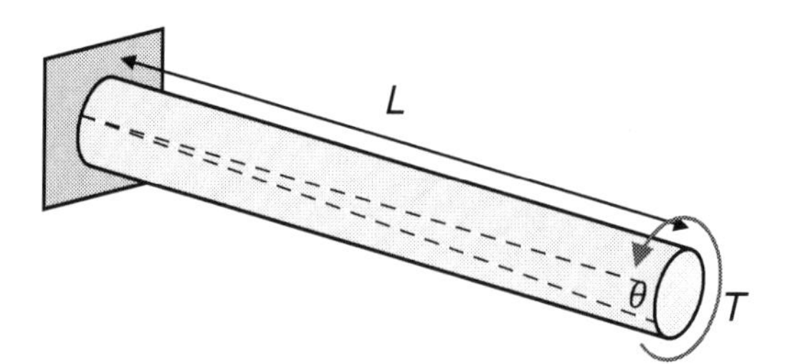

Torsional Stiffness

$$k_t = \frac{T}{\theta} \approx K\frac{G}{L}$$

$k_t = torsional\ stiffness\ [N\ m\ rad^{-1}]$
$K = geometrical\ constant$
$K = J\ for\ circular\ cross-sections$

Section	K (approx.)	Max shear stress
Hollow concentric circle (r_o, r_i)	$\frac{1}{2}\pi(r_o^2 - r_i^2)$	$\frac{2Tr_o}{\pi(r_o^4 - r_i^4)}$
Solid ellipse (2a, 2b)	$\frac{\pi a^3 b^3}{a^2 + b^2}$	$\frac{2T}{\pi ab^2}$ at each end of the minor axis
Solid square (a)	$0.1406a^4$	$\frac{T}{0.208a^3}$
Solid rectangle (a, b)	$\frac{ab^3}{3} \times \left[1 - 0.63\frac{b}{a}\left(1 - \frac{b^4}{12a^4}\right)\right]$	$\frac{T(3a + 1.63)}{a^2b^2} \approx \frac{3T}{ab^2}$ for strip $b \ll a$

4.5. EULER'S BUCKLING CRITERION

Euler Buckling Force

$$F_E = \frac{\pi^2 EI}{(KL)^2}$$

$F_E = Euler\ buckling\ force\ [N]$
$E = Young's\ Modulus\ [Pa]$
$I = Second\ moment\ of\ area\ [m^4]$
$K = column\ effective\ length\ factor$
$L = unsupported\ length\ of\ column\ [m]$
$(KL) = L_e = effective\ length\ [m]$

	End-conditions	K
A	Fixed-Fixed	0.5
B	Fixed-Pinned	0.7
C	Pinned-Pinned	1
D	Fixed-Free	2

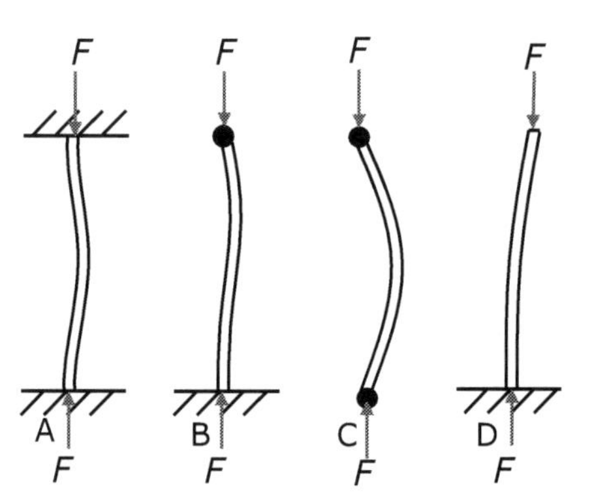

Euler Buckling Stress

$$\sigma_E = \frac{\pi^2 EI}{L_E{}^2 A} = \frac{\pi^2 E}{s^2}$$

$\sigma_E = Euler\ buckling\ stress\ [Pa]$
$I = Ar^2\ [m^4]$
$A = cross-section\ area\ [m^2]$

$$s = \frac{L_E}{R_g}$$

$s = slenderness\ ratio$

$R_g = radius\ of\ gyration\ [m]\ (occasionally\ written\ as\ k\ or\ r_g)$

5. MACHINES AND MECHANISMS

5.1. MECHANISMS

Mobility (Gruebler's Equation)

$$M = 3(n - 1) - 2j_p - j_h$$

$M = degrees\ of\ freedom\ for\ a\ planar\ linkage$
$n = total\ number\ of\ links\ in\ the\ mechanism$
$j_p = total\ number\ of\ primary\ joints\ (pins\ or\ sliding\ joints)$
$j_h = total\ number\ of\ higher-order\ joints\ (cam\ or\ gear\ joints)$

Grashof's Criterion

Grashof's theorem states that a four-bar mechanism has at least one revolving link if:

$$s + l \leq p + q$$

Conversely, the three nonfixed links will merely rock if:

$$s + l > p + q$$

Example: triple rocker $(s + l > p + q)$

$s = length\ of\ shortest\ link$
$l = length\ of\ longest\ link$

$p\ and\ q\ are\ the\ lengths\ of\ intermediate\ links$

Categories of Four-Bar Mechanisms

Case	Criteria	Shortest Link	Category
1	$s + l < p + q$	Frame	Double crank
2	$s + l < p + q$	Side	Crank-rocker
3	$s + l < p + q$	Coupler	Double rocker
4	$s + l = p + q$	Any	Change point
5	$s + l > p + q$	Any	Triple rocker

5.2. CLASSICAL MECHANICS

Linear Terms	Rotary Terms
Force $\vec{F} = m\vec{a}$	Moment (of force) $\vec{M} = \vec{r} \times \vec{F}$
Momentum $p = m\vec{v}$	Angular Momentum $\vec{L} = \vec{r} \times (m\vec{v}) = I\vec{\omega}$
Inertia (=mass) m	Moment of Inertia $I = mk^2$

I = moment of inertia $[kg\ m^2]$. Not to be confused with Second Moment of Area used in Structures.
L = angular momentum L is equivalent to Moment of Momentum H $[kg\ m^2\ s^{-1}]$
k = radius of gyration $[m]$. For a pendulum, this is the distance from axis of rotation ($k = r$).

Derivatives of Linear Position	Derivatives of Angular Position
Average Speed $\bar{v} = \frac{\Delta x}{\Delta t}$	Average Angular Speed $\bar{\omega} = \frac{\Delta \theta}{\Delta t},$
Instantaneous Velocity $\vec{v} = \frac{d\vec{x}}{dt}$	Instantaneous Angular Velocity $\vec{\omega} = \frac{d\vec{\theta}}{dt}$
Velocity Components $v_x = \frac{dx}{dt},\ v_y = \frac{dy}{dt},\ v_z = \frac{dz}{dt}$	Angular Velocity Components $\omega_x = \frac{d\theta_x}{dt},\ \omega_y = \frac{d\theta_y}{dt},\ \omega_z = \frac{d\theta_z}{dt}$
Average Acceleration $\bar{a} = \frac{\Delta v}{\Delta t}$	Average Angular Acceleration $\bar{\alpha} = \frac{\Delta \omega}{\Delta t}$
Instantaneous Acceleration $\vec{a} = \frac{d\vec{v}}{dt}$	Instantaneous Angular Acceleration $\vec{\alpha} = \frac{d\vec{\omega}}{dt}$

$\vec{x}$ = position $[m]$ consisting of components (x, y, z)
$\vec{v}$ = velocity $[m\ s^{-1}]$ consisting of components (v_x, v_y, v_z)
t = time $[s]$
θ = angular position $[rad]$
ω = angular velocity $[rad\ s^{-1}]$
α = angular acceleration $[rad\ s^{-2}]$

Kinematics

Circular Motion

$\vec{v} = \vec{\omega} \times \vec{r}$ VELOCITY

$\vec{a} = \vec{\omega} \times \vec{v} = \vec{\omega} \times (\vec{\omega} \times \vec{r})$ ACCELERATION

$\vec{r} = (linear)\ position\ vector$
$\vec{\omega} = rotation\ vector$

Cartesian Coordinate System

$\vec{r} = x\hat{i} + y\hat{j} + z\hat{k}$

$\vec{v} = v_x\hat{i} + v_y\hat{j} + v_z\hat{k}$

$\vec{a} = a_x\hat{i} + a_y\hat{j} + a_z\hat{k}$

Acceleration of a point in two dimensions can contain radial, centripetal, tangential and Coriolis components. Cylindrical coordinates introduce an additional component in the $\hat{k}$ direction.

Rotating Coordinate System

$\vec{x} = r\hat{e}_r + z\hat{e}_z$ POSITION

$\vec{v} = \dot{r}\hat{e}_r + r\omega\hat{e}_\theta + \dot{z}\hat{e}_z$ VELOCITY

$\vec{a} = (\ddot{r} - r\omega^2)\hat{e}_r + (r\alpha + 2\dot{r}\omega)\hat{e}_\theta + \ddot{z}\hat{e}_z$ ACCELERATION

Unit vector $\hat{e}_z$ points along the direction of the chosen reference axis (the axis of rotation). Unit vectors $\hat{e}_r$ and $\hat{e}_\theta$ form the **radial** and **tangential** directions of a point, where $\hat{e}_r$ points from the reference axis, and $\hat{e}_\theta$ points at 90° to both $\hat{e}_r$ and $\hat{e}_z$.

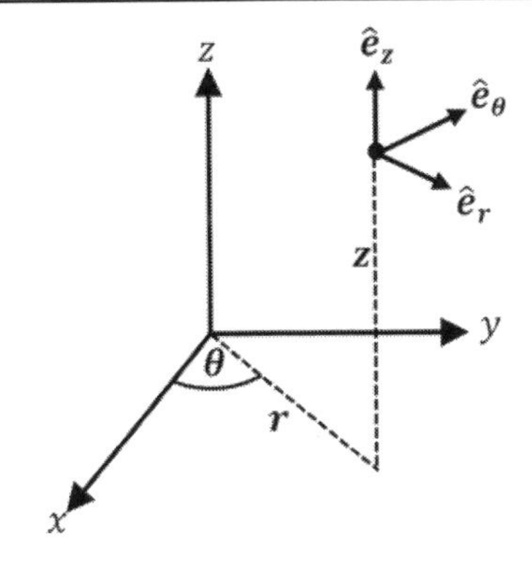

Complex Notation

Planar motion may be described using complex numbers.

Planar Motion of a Point in C-Space

The position of a point in the complex plane is given in terms of Euler's notation:

$$x = re^{j\theta}$$

Components of Velocity

$$v = \frac{d}{dt}(re^{j\theta}) = (\dot{r} + jr\omega)e^{j\theta}$$

$\dot{r}e^{j\theta} = radial\ velocity$
$jr\omega e^{j\theta} = tangential\ velocity$

Components of Acceleration

$$a = \frac{d^2}{dt^2}(re^{j\theta}) = \frac{d}{dt}(\dot{r}e^{j\theta} + jr\omega e^{j\theta})$$

$$= \left[(\ddot{r} - r\omega^2) + j(r\alpha + 2\dot{r}\omega)\right]e^{j\theta}$$

$\ddot{r}e^{j\theta} = radial\ acceleration$
$-r\omega^2 e^{j\theta} = centripetal\ acceleration$
$r\alpha e^{j\theta} = tangential\ acceleration$
$2\dot{r}\omega e^{j\theta} = coriolis\ acceleration$

The imaginary unit j represents a 90° anticlockwise rotation.
The term $e^{j\theta}$ represents a rotation of θ radians anticlockwise.
Radial acceleration is in the direction away from the origin.
Centripetal acceleration is in the direction towards the origin.
Tangential acceleration and **Coriolis acceleration** are at 90° from the line connecting the point x to the origin.

Equations of Motion (Constant Acceleration)

Translation	Rotation
Final position:	Final angular position:
$x = x_0 + v_0 t + \frac{1}{2}at^2$	$\theta = \theta_0 + \omega_0 t + \frac{1}{2}\alpha t^2$
$x = x_0 + \left(\frac{v_0 + v}{2}\right)t = x_0 + \bar{v}t$	$\theta = \theta_0 + \left(\frac{\omega_0 + \omega}{2}\right)t = \theta_0 + \bar{\omega}t$
Final speed:	Final angular speed:
$v = v_0 + at$	$\omega = \omega_0 + \alpha t$
$v = \sqrt{v_0^2 + 2a(x - x_0)}$	$\omega = \sqrt{\omega_0^2 + 2\alpha(\theta - \theta_0)}$
Average speed:	Average angular speed:
$\bar{v} = \frac{1}{2}(v + v_0)$	$\bar{\omega} = \frac{1}{2}(\omega + \omega_0)$
$x_0 = initial\ position$ $v_0 = initial\ velocity$ $a = a_0 = acceleration$	$\theta_0 = initial\ angular\ position\ (rad)$ $\omega_0 = initial\ angular\ velocity\ (rad\ s^{-1})$ $\alpha = \alpha_0 = angular\ acceleration\ (rad\ s^{-2})$
Velocity – Time Equation	Velocity – Time Equation
$\int_{v_0}^{v} dv = \int_0^t a dt$	$\int_{\omega_0}^{\omega} d\omega = \int_0^t \alpha dt$
Position – Time Equation	Position – Time Equation
$\int_{s_0}^{s} ds = \int_0^t (v_0 + at) dt$	$\int_{\theta_0}^{\theta} d\theta = \int_0^t (\omega_0 + \alpha t) dt$
Velocity – Position Equation	Velocity – Position Equation
$\int_{v_0}^{v} dv = \int_{s_0}^{s} a ds$	$\int_{\omega_0}^{\omega} d\omega = \int_{\theta_0}^{\theta} \alpha d\theta$

Kinetics

Newton's Second Law

Force is equal to the rate of change of linear momentum:

$$\sum \vec{F} = \frac{d\vec{p}}{dt} = m\frac{d\vec{v}}{dt} + \vec{v}\frac{dm}{dt}$$

For a constant mass of particles, the net force is given by the sum of forces acting upon each particle:

$$\sum \vec{F} = m\frac{d\vec{v}}{dt} = m\vec{a}$$

Impulse

$$\vec{J} = \int_{t_1}^{t_2} \vec{F}dt = \int_{t_1}^{t_2} (m\vec{a})dt = \int_{t_1}^{t_2} md\vec{v} = [m\vec{v}]_{t_1}^{t_2} = \Delta(m\vec{v})$$

An impulse is the integral of force over a time interval. An impulse applied to an object produces an equivalent vector change in its linear momentum. Examples of applications of the concept of impulse include shock absorbers, air bags, braking force, rocket motors, high-speed collisions and calculations involving projectiles.

The average applied force is given by the change in impulse over time:

$$\sum \vec{F} = m\left(\frac{v_2 - v_1}{\Delta t}\right)$$

Torque

$$\tau = rF\sin\,\theta \quad or \quad \vec{\tau} = \vec{r} \times \vec{F}$$

Angular Momentum

$$\vec{L} = I\vec{\omega}$$

$I = moment\ of\ inertia\ [kg\,m^2]$
$\omega = angular\ speed\ [rad\,s^{-1}]$

$$\vec{L} = \vec{r} \times \vec{p} = \vec{r} \times (m\vec{v})$$

$\vec{r} = (linear)\ position\ vector$
$\vec{p} = momentum\ vector$

Radius of Gyration (Kinetics)

$$k_{axis} = \sqrt{\frac{I_{axis}}{m}}$$

Moment of Inertia (Kinetics)

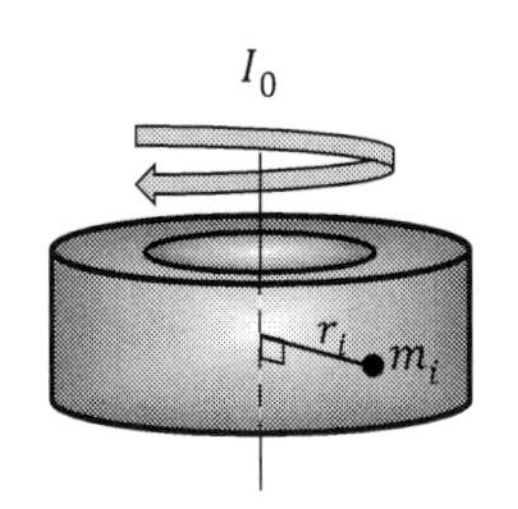

$$I = \sum (m_i r_i^2)$$

$$I = \int r^2 dm = mk^2$$

$m = total\ mass\ [kg]$
$k = radius\ of\ gyration\ [m]$

Power

Translation	Rotation
Linear power $P = \vec{F} \cdot \vec{v}$ $\vec{F} = force\ [N]$ $\vec{v} = velocity$	Rotating power $P = \vec{\tau} \cdot \vec{\omega}$ $\tau = torque\ [N\ m]$ $\omega = rotational\ speed\ [rad\ s^{-1}]$
Average linear power $\bar{P} = \bar{F} v \cos\theta$ $\bar{F} = average\ force$ $\theta = angle\ between\ vectors$	Average rotating power $\bar{P} = \bar{\tau} \omega \cos\theta$ $\bar{\tau} = average\ torque$ $\theta = angle\ between\ vectors$

Kinetic Energy

Linear kinetic energy	Rotational kinetic energy
$T = \frac{1}{2} m v^2$	$T = \frac{1}{2} I \omega^2$

Potential Energy

$\Delta V = -\int F \cdot ds$ GENERAL POTENTIAL ENERGY

$\Delta V_g = mg\Delta h$ GRAVITATIONAL POTENTIAL ENERGY

$\Delta V_s = \frac{1}{2} k \Delta x^2$ ELASTIC (SPRING) POTENTIAL ENERGY

Uniform Circular Motion

$$F = \frac{mv^2}{r} = mr\omega^2 \quad \text{CENTRIPETAL FORCE}$$

$$a = \frac{2\pi v}{T} = \frac{v^2}{r} = r\omega^2 \quad \text{ACCELERATION}$$

$$\omega = \frac{v}{r} = 2\pi f = \frac{2\pi}{T} \quad \text{ANGULAR SPEED}$$

$$T = \frac{2\pi r}{v} \quad \text{PERIOD}$$

$F = centripetal\ force\ [N]$
$m = point\ mass\ [kg]$
$a = (linear)\ acceleration\ [m\,s^{-2}]$
$v = (linear)\ speed\ [m\,s^{-1}]$

$\omega = angular\ speed\ [rad\,s^{-1}]$
$r = radius\ [m]$
$f = frequency\ of\ rotation\ [Hz]$
$T = period\ of\ one\ rotation\ [s]$

Friction

$$F \leq \mu N$$

$F = friction\ force\ [N]$
$\mu = dry\ friction\ coefficient$
$N = normal\ force\ [N]$

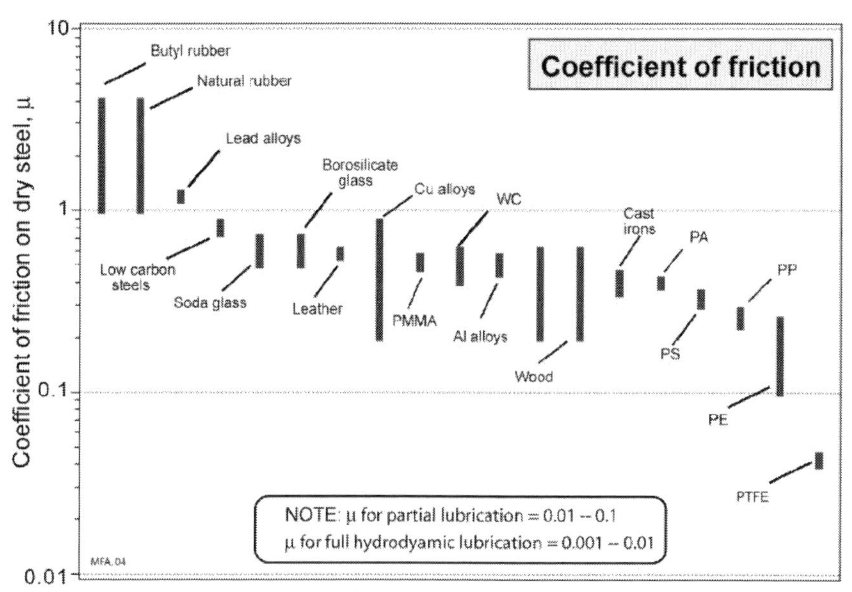

Friction coefficient is typically around 0.5

Materials property chart courtesy of Granta Design, www.grantadesign.com

5.3. PERIODIC MOTION

Springs (Hooke's Law)

$$F = kx = \left(\frac{1}{C}\right)x$$

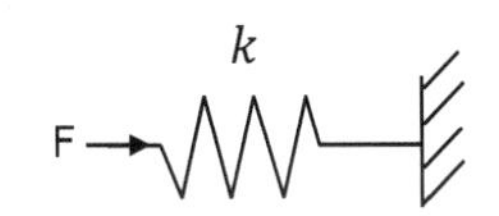

$k = spring\ stiffness\ [N\,m^{-1}]$
$C = spring\ compliance\ [m\,N^{-1}]$

Dampers or *Dashpots*

$$F = c\frac{dx}{dt}$$

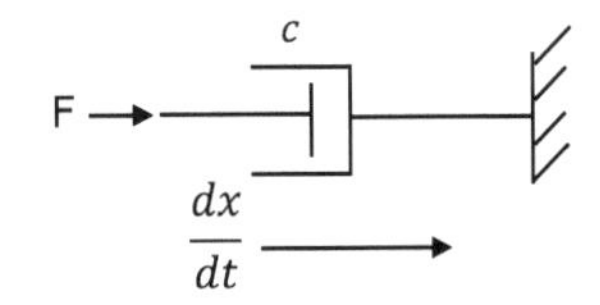

$c = damping\ coefficient\ [kg\,s^{-1}]$

Simple Harmonic Motion

Equation of motion (undamped)

$$\ddot{x} + {\omega_n}^2 x = 0$$

Solution

$$x(t) = A\cos(\omega t - \varphi)$$

Frequency

$$f = \frac{1}{T}$$

$$\omega = 2\pi f$$

$T = period\ (s)$

$f = frequency\ (Hz)$

$\omega = angular\ frequency\ (rad\ s^1)$

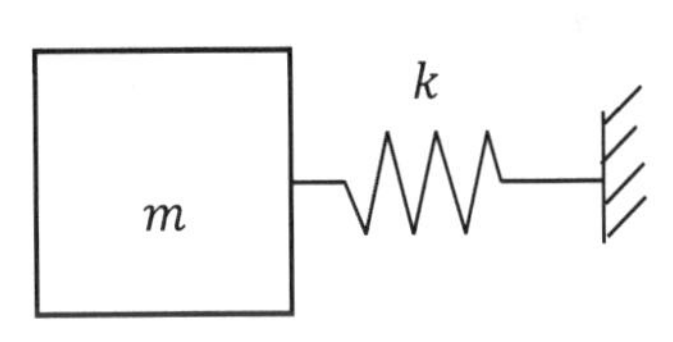

Mass-Spring-Damper

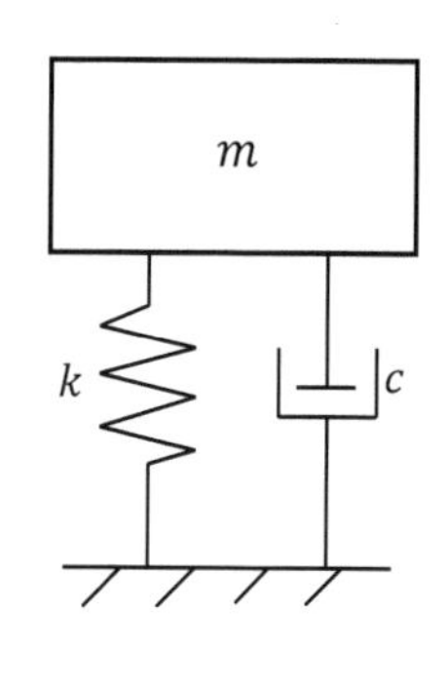

Equation of motion (damped)

$$m\ddot{x} + c\dot{x} + kx = 0$$

or

$$\ddot{x} + 2\zeta\omega_n\dot{x} + {\omega_n}^2 x = 0$$

Damping ratio:

$$\zeta = \frac{c}{2\sqrt{mk}} = \frac{c}{2m\omega_n}$$

$c = viscous\ damping\ coefficient$

$$\omega_n = \sqrt{\frac{k}{m}}$$ UNDAMPED NATURAL FREQUENCY

$$T = 2\pi\sqrt{\frac{m}{k}}$$ PERIOD OF OSCILLATION

Simple Pendulum

$$\omega_n = \sqrt{\frac{g}{l}}$$

$$T = 2\pi\sqrt{\frac{l}{g}}$$

$\omega_n = natural\ frequency$
$T = period\ [s]$

5.4. LAGRANGE'S EQUATION

$$L = T - V$$ THE LAGRANGIAN

$$\frac{d}{dt}\left(\frac{\partial L}{\partial \dot{q}_j}\right) - \frac{\partial L}{\partial q_j} = 0$$

$T = total\ kinetic\ energy$
$V = total\ potential\ energy$
$q_j = generalised\ coordinate\ for\ the\ j^{th}\ degree\ of\ freedom$

> Lagrange's Equation gives the equations of motion for a conservative system, with one equation for each generalised coordinate.

6. ELECTRICAL AND ELECTRONICS

6.1. FUNDAMENTALS OF ELECTROMAGNETISM

Ohm's Law

$v = iR$ INSTANTANEOUS VOLTAGE

$i = \frac{dq}{dt}$ INSTANTANEOUS CURRENT

$P = iv = i^2 R$ POWER LOSS

$v = instantaneous\ voltage\ [V]$
$i = instantaneous\ current\ [A]$
$R = resistance\ [\Omega]$
$P = DC\ power\ [W]$
$q = charge\ [C]$
$t = time\ [s]$

Resistance

$$R = \frac{\rho l}{A} = \int_0^l \frac{\rho_0(1 + \alpha T)}{A} dx$$

$\rho = resistivity\ of\ material\ (\Omega\ m)$
$\rho_0 = resistivity\ at\ temperature\ 0°C\ (\Omega\ m)$
$T = temperature\ (°C)$
$A = cross\ section\ area\ of\ conductor\ (m^2)$
$l = length\ of\ conductor\ (m)$
$\alpha = resistance\ thermal\ coefficient$

Inductance

$$v = L\frac{di}{dt} = L\frac{d^2q}{dt^2}$$

$$L = \frac{N^2 \mu_0 \mu_r A}{l}$$

$L = inductance\ [H]$
$\mu_0 = vacuum\ permeability = 4\pi \times 10^{-7}\ H\ m^{-1}$
$\mu_r = relative\ permeability$
$N = number\ of\ turns\ in\ inductive\ coil$

Capacitance

$q = CV = \int i\,dt$ STORED CHARGE

$i = C\frac{dv}{dt}$ INSTANTANEOUS CURRENT

$C = \varepsilon_0\varepsilon_r(n-1)A/d$ PARALLEL PLATES

$C = capacitance\ [F]$
$\varepsilon_0 = vacuum\ permittivity\ (8.854 \times 10^{-12}\ F\,m^{-1})$
$\varepsilon_r = relative\ permittivity$
$n = number\ of\ parallel\ plates$
$A = area\ of\ plates\ [m^2]$
$d = distance\ between\ plates\ [m]$

Stored Energy

$$E_L = \frac{1}{2}Li^2$$

$$E_C = \frac{1}{2}Cv^2$$

$E_L = energy\ in\ an\ inductor\ [J]$
$E_C = energy\ in\ a\ capacitor\ [J]$

Reactance

$X_L = \omega L = 2\pi f L$ INDUCTIVE REACTANCE

$X_C = \frac{-1}{\omega C}$ CAPACITIVE REACTANCE

$X_C = capacitive\ reactance\ [\Omega]$
$X_L = inductive\ reactance\ [\Omega]$
$f = cyclic\ frequency\ [Hz]$
$\omega = angular\ frequency\ [rad\ s^{-1}]$
$L = inductance\ [H]$

$C = capacitance\ [F]$

Complex Impedance

$$\tilde{V} = \tilde{I}Z$$

$$Z = R + jX$$

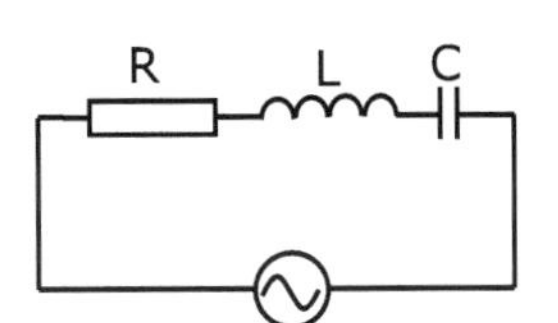

$Z = impedance$
$R = resistance$
$X = total\ reactance$

$$Z_R = R$$

$$Z_L = j\omega L$$

$$Z_C = \frac{1}{j\omega C} = -\frac{j}{\omega C}$$

$$|Z| = \sqrt{R^2 + (X_L + X_C)^2}$$

$|Z| = magnitude\ of\ impedence\ [\Omega]$

Parallel impedances

$$Z_t = \frac{z_1 z_2}{z_1 + z_2}$$

Series impedances

$$Z_{total} = (R + Z_L + Z_C) = R + j\left(\omega L - \frac{1}{\omega C}\right)$$

Electrical Resonance (series circuit)

$$X_L = X_C$$

$$2\pi f L = \frac{1}{2\pi f C} \Rightarrow f = \sqrt{\frac{1}{4\pi^2 LC}}$$

$$f_r = \frac{1}{2\pi\sqrt{LC}}$$

$$\omega_r = \frac{1}{\sqrt{LC}}$$

$f_r = cyclic\ frequency\ [Hz]$
$\omega_r = angular\ frequency\ of\ resonance\ [rad\ s^{-1}]$

Electrical resonance occurs in an AC circuit when equal and opposite reactances cancel each other out i.e. $X_L = X_C$.

6.2. TRANSFORMERS

Induced EMF in a coil

$$E = - N\frac{d\phi}{dt}$$

$N = number\ of\ turns$
$\phi = flux\ [webers]$

$$\phi = BA\cos\theta$$

$\phi = change\ in\ magnetic\ flux\ through\ a\ rotating\ loop\ [T \cdot m^2]$
$B = magnetic\ field\ strength\ [T]$
$A = area\ [m^2]\ at\ angle\ \theta\ to\ the\ magnetic\ flux$

Ideal Transformers

Induced electromotive force per turn is constant across a transformer:

$$\frac{E_1}{N_1} = \frac{E_2}{N_2}$$

$E_1 = V_1$
$E_2 = V_2$

$$\frac{V_1}{V_2} = \frac{N_1}{N_2} = \frac{I_2}{I_1}$$

I_1 I_2 V_1 N_1 N_2 V_2

Induced EMF in a transformer

$$E = \sqrt{2}\pi f N\phi \approx 4.44 f N\phi_{pk}$$

$f = frequency\ of\ AC\ supply\ [Hz]$
$N = number\ of\ turns\ in\ a\ winding$
$\phi_{pk} = peak\ value\ of\ magnetic\ flux\ [T]$

% voltage regulation

$$Percent\ VR = \frac{|V_{OC}| - |V_{load}|}{|V_{OC}|} \times 100\%$$

$V_{OC} = open\ circuit\ (no\ load)\ voltage$
$V_{load} = full\ load\ voltage$

6.3. ELECTRICAL MACHINES

Lorentz Force

$\vec{F} = q(\vec{E} + \vec{v} \times \vec{B})$ FORCE ON A CHARGE IN EM FIELD

$B = \frac{\mu_o I}{2\pi R}$ MAGNETIC FIELD AROUND A WIRE

DC Machines

$E = vBl$ EMF IN A MOVING WIRE

$E = e.m.f.\ [V]$
$v = velocity\ [m\ s^{-1}]$
$B = flux\ density\ [T]$
$l = length\ of\ conductor\ [m]$

$F = iBl$ FORCE ON A CONDUCTING WIRE

$F = force\ [N]$
$i = instantaneous\ current$

$E = k_a \phi \omega \left(where\ k_a = \frac{zp}{2\pi a} \right)$ EMF OF A DC MACHINE

$\omega = rotational\ frequency\ [rad\ s^{-1}]$
$p = number\ of\ poles$
$k = proportionality\ constant$
$a = 2\ for\ wave\ winding$
$a = p\ for\ lap\ winding$

Electromechanical conversion:

$$T = \frac{pZ}{2\pi A} \phi I_a$$

$$k = \frac{pZ}{2\pi A}$$

$$T = k_a \phi I_a$$

$p = number\ of\ poles$
$Z = total\ no.\ of\ armature\ conductors = no.\ of\ slots \times no.\ of\ conductors/slot$
$A = no.\ of\ parallel\ paths\ in\ armature$
$T = generated\ torque\ [N\ m]$
$k_a = proportionality\ constant$
$\phi = flux\ per\ pole\ [Wb]$
$I_a = current\ [A]$

Shunt machine

$$V_{motor} = E + I_a R_a$$

$$V_{generator} = E - I_a R_a$$

V = terminal voltage (for motor or generator)
E = back e.m.f.
$I_a R_a$ = armature losses

Series machine

$$V_{motor} = E + I_a(R_a + R_f)$$

$$V_{generator} = E - I_a(R_a + R_f)$$

V = terminal voltage [V]
I_a = armature current [A]
R_a = armature resistance [Ω]
R_f = field resistance [Ω]

AC Machines

$$\omega_s = \frac{2}{p} 2\pi f$$

$$\omega = \omega_s(1 - s)$$

ω_s = synchronous speed $[rad\ s^{-1}]$
f = AC frequency [Hz]
s = slip

6.4. AC POWER

Single Phase

$$V_{rms} = \frac{1}{\sqrt{2}}(V_{peak})$$

$$V_{average} = \frac{2}{\pi}(V_{peak}) = \frac{2\sqrt{2}}{\pi}(V_{rms})$$

$S = I_{rms}V_{rms}$ APPARENT POWER

$P = S\cos\varphi = I_{rms}V_{rms}\cos\varphi$ REAL POWER

$Q = S\sin\varphi = I_{rms}V_{rms}\sin\varphi$ REACTIVE POWER

$\cos\varphi = \frac{P}{|S|} = \frac{R}{|Z|}$ POWER FACTOR

$V_{rms} = RMS\ voltage$
$I = RMS\ current$
$P = real\ power\ [kW]$
$S = apparent\ power\ [VA]$
$Q = reactive\ power\ [VAR = volt \cdot amps\ reactive]$
$R = circuit\ resistance\ [\Omega]$
$Z = circuit\ impedance\ [\Omega]$

Balanced 3-Phase

$$Total\ power = 3\,V_{ph}I_{ph}\cos\varphi$$

$V_{ph} = RMS\ voltage\ across\ one\ phase$
$I_{ph} = RMS\ current\ across\ one\ phase$
$\varphi = angle\ between\ phases$

$$Total\ power = \sqrt{3}\,V_{L}I_{L}\cos\varphi$$

$V_{L} = RMS\ voltage\ between\ lines$
$I_{L} = RMS\ current\ in\ a\ line$

Star-delta Conversions

Delta to Star

$$z_1 = \frac{z_{12}z_{31}}{\Sigma}$$

$$z_2 = \frac{z_{23}z_{12}}{\Sigma}$$

$$z_3 = \frac{z_{31}z_{23}}{\Sigma}$$

$$\Sigma = z_1 + z_2 + z_3$$

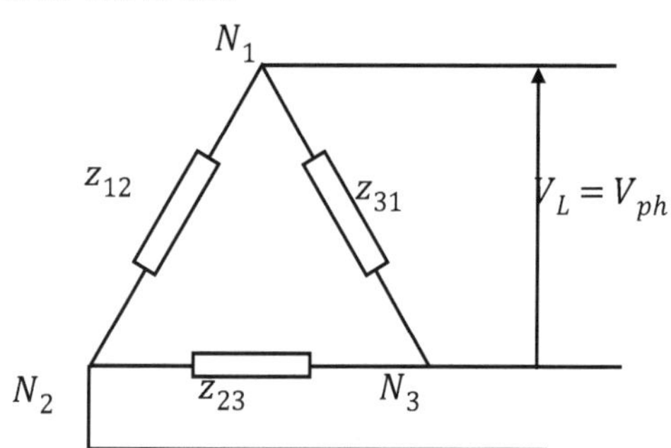

Star to Delta

$$z_{12} = \frac{P}{z_3}$$

$$z_{23} = \frac{P}{z_1}$$

$$z_{31} = \frac{P}{z_2}$$

$$P = z_1z_2 + z_2z_3 + z_3z_1$$

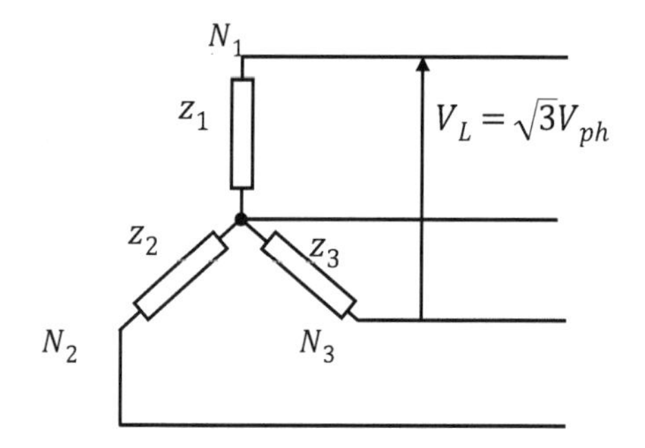

Complex Waveforms

$$v = V_1\sin(\omega t + \varphi_1) \pm V_2\sin(2\omega t + \varphi_2) \pm V_3\sin(3\omega t + \varphi_3)... \pm V_n\sin(n\omega t + \varphi_n)$$

$v = instantaneous\ voltage\ at\ time\ t$
$V_1 = peak\ amplitude\ of\ fundamental\ harmonic\ (= \sqrt{2}\cdot V_{rms})$
$V_n = amplitude\ of\ n^{th}\ harmonic$
$\frac{\omega}{2\pi} = fundamental\ frequency$
$\varphi_n = phase\ angle\ of\ the\ n^{th}\ harmonic\ relative\ to\ fundamental$

Any continuous function can be represented as a sum of sine or cosines, each with the form $v_i = V_i\sin(k\omega t + \varphi_i)$. This is known as a Fourier series.

6.5. ELECTRONICS

Circuit Diagram Symbols

	Joined conductors		Crossing conductors – no connection		Single-Pole-Single-Throw switch (SPST) normally open)
	Fixed resistor		Diode		Single-Pole-Single-Throw switch (SPST) (normally closed)
	Potentiometer		Light-Emitting-Diode (LED)		Single-Pole-Double-Throw switch (SPDT)
	preset potentiometer		Zener diode		Double-Pole-Double-Throw switch (DPDT)
	Thermistor		Schottky diode		Push-To-Make switch (PTM)
	Light-dependent resistor		Amplifier		Push-To-Break switch (PTB)
	Polarised capacitor		Fuse		Dry-reed switch
	Non-polarised capacitor	2 pin	Resonator		Opto switch
	Inductor	3 pin			NPN transistor
e.g. +9V 0V	Power supply		Primary or secondary cell	RL	Relay (with double-throw contacts – contact symbol varies with type used)
			Battery (of cells)		

Diodes

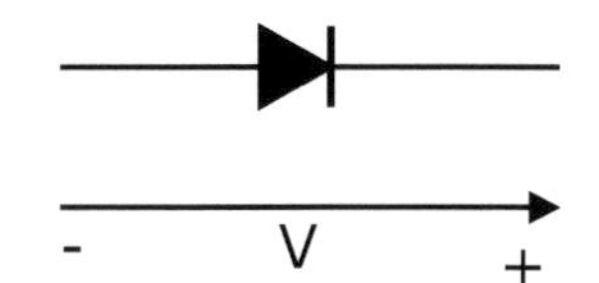

$$I = I_S\left(exp\frac{eV}{\eta kT} - 1\right)$$

For most practical purposes

$$I \approx I_S\left(exp\frac{eV}{kT} - 1\right)$$

$I = diode\ current$
$I_S = reverse\ saturation\ current$
$e = electronic\ charge$
$V = diode\ voltage$
$\eta = material\ constant$
$k = Boltzmann\ constant$
$T = absolute\ temperature$

Field Effect Transistors

MOSFET

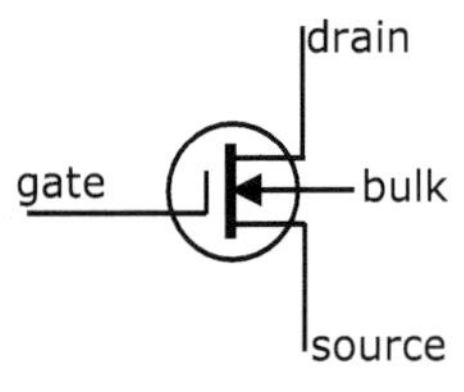

n-channel DE

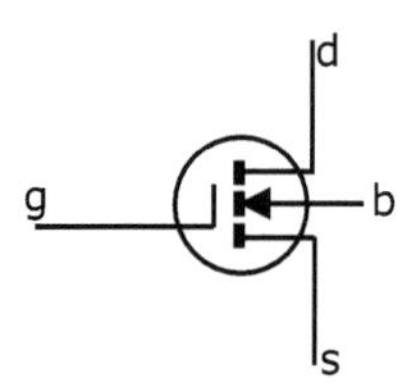

n-channel enhancement

$I_D = K(V_{GS} - V_T)^2$

$I_D = drain\ current$

$V_{GS} = gate\ to\ source\ voltage$

$V_T = threshold\ voltage$

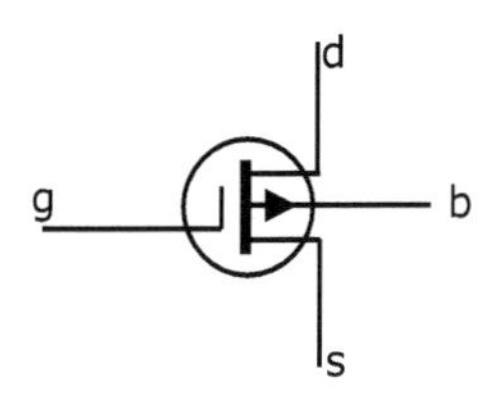

p-channel DE

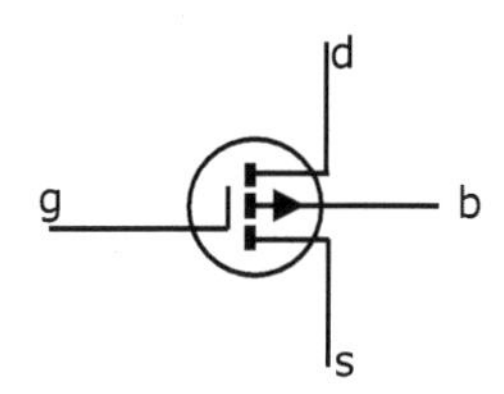

p-channel enhancement

JFET

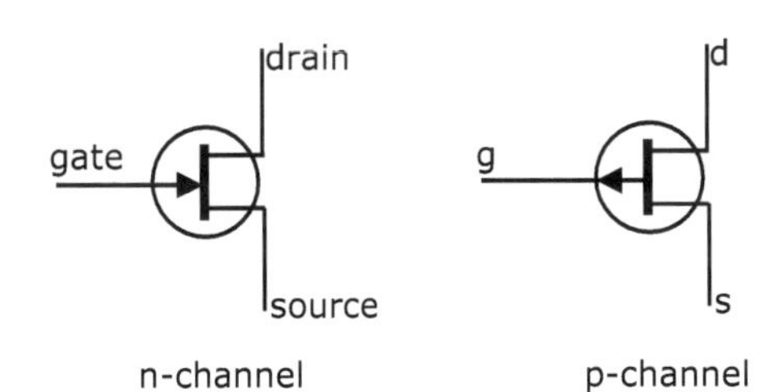

n-channel p-channel

$$I_D = I_{DSS}\left(1 - \frac{V_{GS}}{V_P}\right)^2$$

$$= k'(V_{GS} - V_P)^2$$

$I_D = drain\ current$

$V_{DSS} = drain\ to\ source\ saturation\ current$

$V_{GS} = gate\ to\ source\ voltage$

$V_P = pinch\ off\ voltage$

Bipolar Junction Transistors

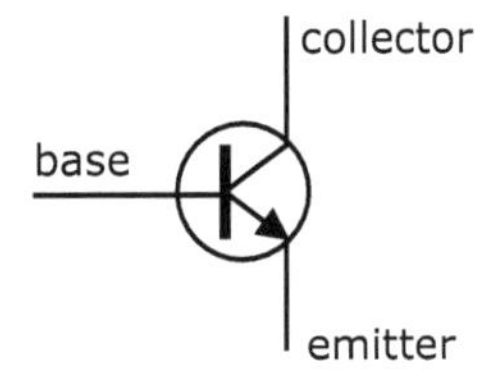

c

b

e

npn transistor pnp transistor

$$g_m = \frac{dI_C}{dV_{BE}} \approx 40I_E$$

$$I_B = I_{BS}\left(\exp\left(\frac{eV_{BE}}{kT}\right) - 1\right)$$

$$I_C = h_{FE}I_B \qquad i_c = h_{fe}i_b$$

$g_m = transconductance$
$I_C = collector\ current$
$V_{BE} = base\ to\ emitter\ voltage$
$I_E = emitter\ current$

$I_B = base\ current$
$I_{BS} = base\ saturation\ current$
$e = electronic\ charge$
$k = Boltzmann\ constant$
$T = absolute\ temperature$

$i_c = small\ signal\ collector\ current$
$i_b = small\ signal\ base\ current$
$h_{FE} = DC\ current\ gain$
$h_{fe} = AC\ or\ small\ signal\ current\ gain$

Bipolar Amplifier Configurations

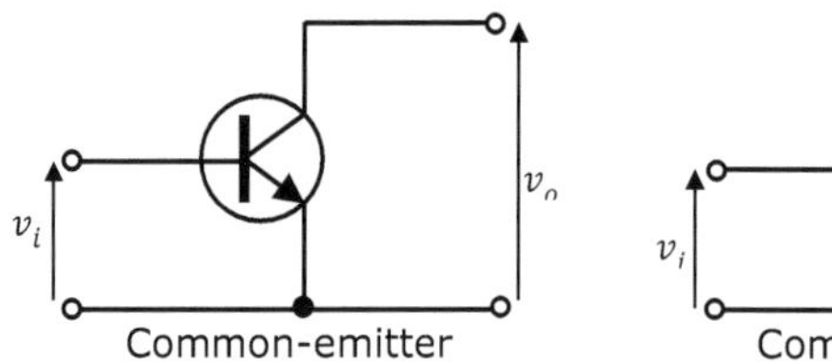

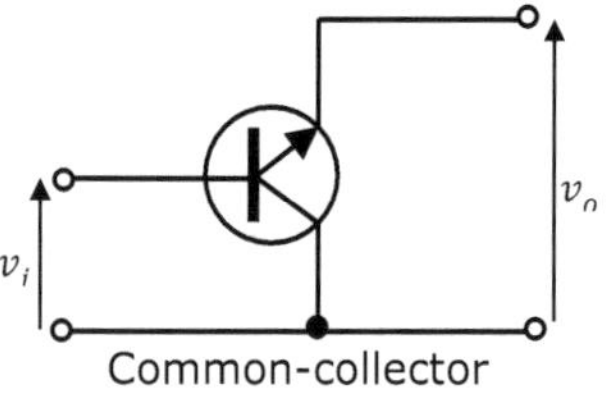

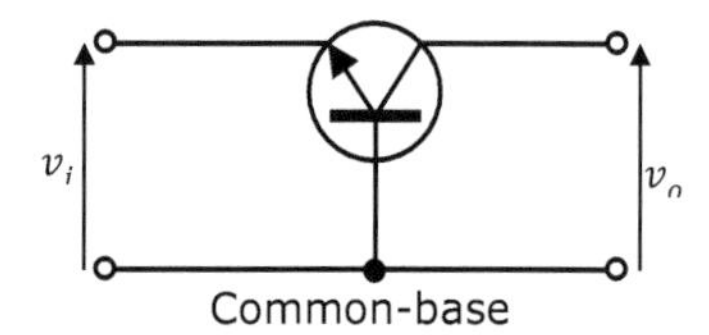

Table of Amplifier Configurations

	Common-emitter	Common-collector	Common-base
Input terminal	base	base	emitter
Output terminal	collector	emitter	collector
Voltage gain A_v	$-\,g_m R_C\,(high)$	$\approx 1\,(unity)$	$g_m R_C\,(high)$
Current gain A_i	$-\,h_{fe}\,(high)$	$h_{fe}\,(high)$	≈ -1
Power gain A_p	$A_v A_i\,(very\ high)$	$\approx A_i\,(high)$	$\approx A_v\,(high)$
Input impedance	$R_1//R_2$ $(moderate)$	$R_1//R_2$ $(moderate)$	$\approx r_e\,(very\ low)$
Output impedance	$R_c/\frac{1}{h_{oe}}$ $(moderate)$	$\approx r_e\,(very\ low)$	$\approx R_C\,(high)$
Phase-shift (mid-band)	180°	0°	0°

Operational Amplifiers

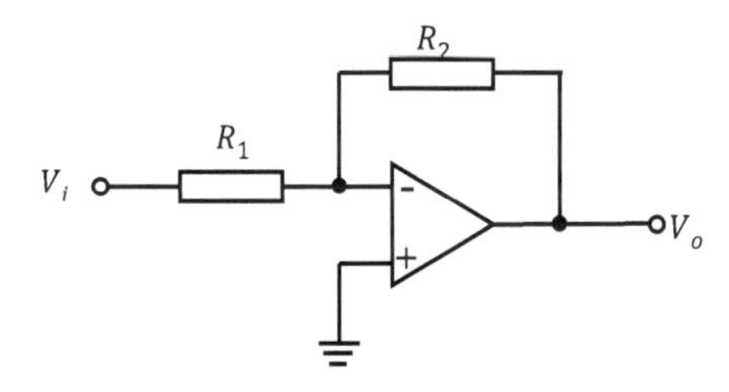

Inverting amplifier

$$V_o = -\frac{R_2}{R_1} V_i$$

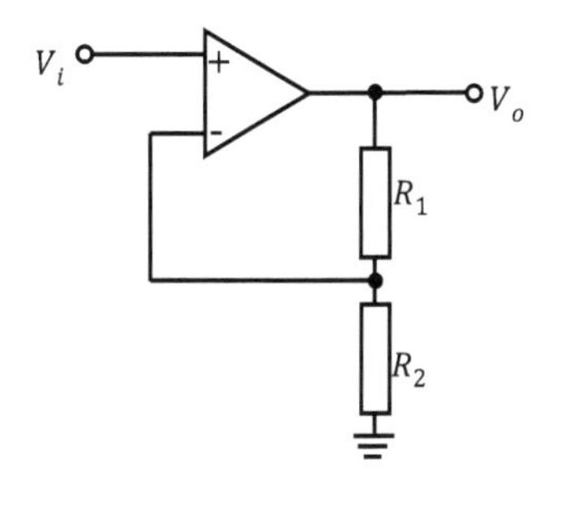

Non-inverting amplifier

$$V_o = \frac{R_1 + R_2}{R_2} V_i$$

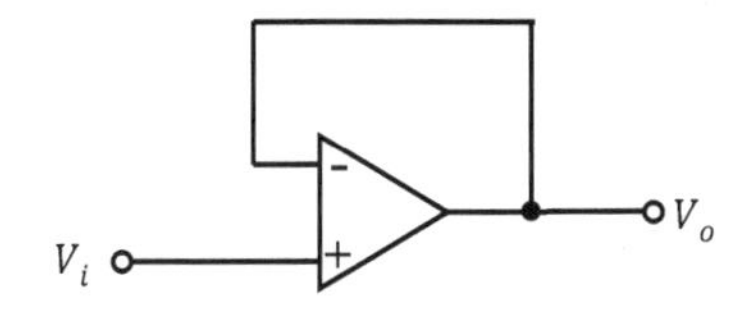

Unity gain buffer amplifier

$$V_o = V_i$$

Logic Gates

Function	Symbol	Boolean	Truth Table
AND (intersection)	A, B → C	$C = A \cdot B$	A B | C 0 0 | 0 0 1 | 0 1 0 | 0 1 1 | 1
OR (union)	A, B → C	$C = A + B$	A B | C 0 0 | 0 0 1 | 1 1 0 | 1 1 1 | 1
NOT	A → B	$B = \overline{A}$	A | B 0 | 1 1 | 0
NAND	A, B → C	$C = \overline{A \cdot B}$	A B | C 0 0 | 1 0 1 | 1 1 0 | 1 1 1 | 0
NOR	A, B → C	$C = \overline{A + B}$	A B | C 0 0 | 1 0 1 | 0 1 0 | 0 1 1 | 0
Exclusive OR	A, B → C	$C = A \oplus B$	A B | C 0 0 | 0 0 1 | 1 1 0 | 1 1 1 | 0
Exclusive NOR	A, B → C	$C = \overline{A \oplus B}$	A B | C 0 0 | 1 0 1 | 0 1 0 | 0 1 1 | 1

Boolean Algebra

Identities and Laws

AND Function

(1) $0 \cdot 0 = 0$
(2) $0 \cdot 1 = 0$
(3) $1 \cdot 0 = 0$
(4) $1 \cdot 1 = 1$
(5) $A \cdot 0 = 0$
(6) $0 \cdot A = 0$
(7) $A \cdot 1 = A$
(8) $1 \cdot A = A$
(9) $A \cdot A = A$
(10) $A \cdot \bar{A} = 0$

OR Function

(11) $0 + 0 = 0$
(12) $0 + 1 = 1$
(13) $1 + 0 = 1$
(14) $1 + 1 = 1$
(15) $A + 0 = A$
(16) $0 + A = A$
(17) $A + 1 = 1$
(18) $1 + A = 1$
(19) $A + A = A$
(20) $A + \bar{A} = 1$

NOT Function

(21) $\bar{0} = 1$
(22) $\bar{1} = 0$
(23) $\bar{\bar{A}} = A$

Commutative Law

(24) $AB = BA$
(25) $A + B = B + A$

Distributive Law

(26) $A(B + C) = AB + AC$
(27) $A + BC = (A + B)(A + C)$

Associative Law

(28) $A(BC) = (AB)C$
(29) $A + (B + C) = (A + B) + C$

Absorption Law

(30) $A + AB = A$
(31) $A(A + B) = A$

DeMorgan's Law

(32) $\overline{A + B} = \bar{A} \cdot \bar{B}$
(33) $\overline{A \cdot B} = \bar{A} + \bar{B}$

Also note

(34) $A + \bar{A}B = A + B$
(35) $A(\bar{A} + B) = AB$

Resistor Colour Code Table

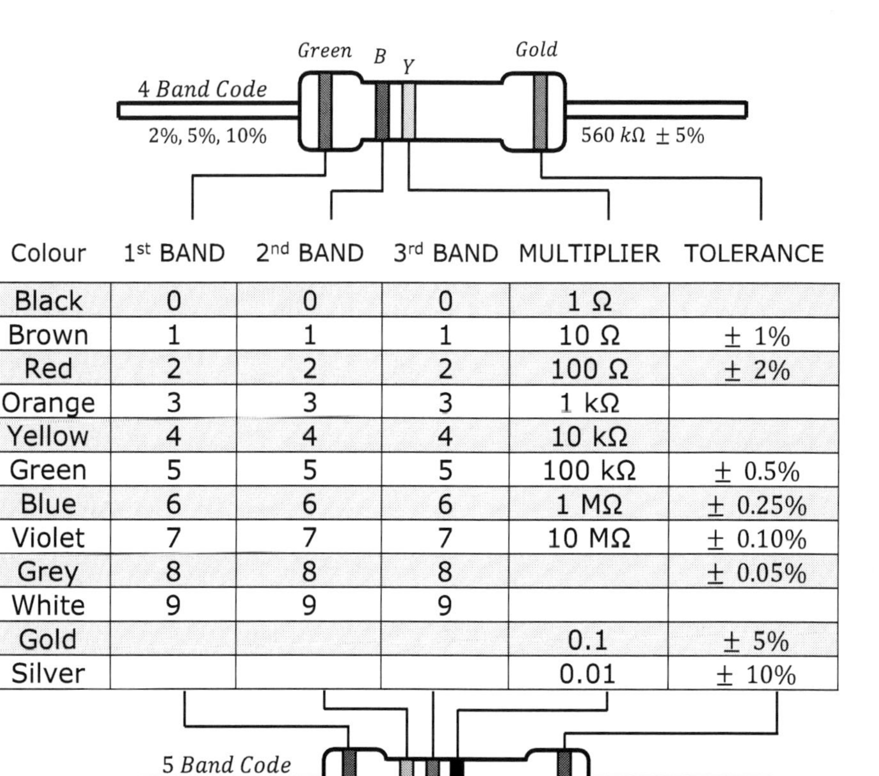

Colour	1st BAND	2nd BAND	3rd BAND	MULTIPLIER	TOLERANCE
Black	0	0	0	1 Ω	
Brown	1	1	1	10 Ω	± 1%
Red	2	2	2	100 Ω	± 2%
Orange	3	3	3	1 kΩ	
Yellow	4	4	4	10 kΩ	
Green	5	5	5	100 kΩ	± 0.5%
Blue	6	6	6	1 MΩ	± 0.25%
Violet	7	7	7	10 MΩ	± 0.10%
Grey	8	8	8		± 0.05%
White	9	9	9		
Gold				0.1	± 5%
Silver				0.01	± 10%

7. THERMODYNAMICS

7.1. FIRST LAW

$$\Sigma Q_{in} - \Sigma W_{out} = \Delta E$$

Net heat into a closed system equals net work output plus change in **total energy**, which may include internal energy, kinetic energy, or potential energy for example.

$$\Sigma Q_{in} - \Sigma W_{out} = \Delta U + \Delta KE + \Delta PE$$

$\Sigma Q_{in} = (Q_{in} - Q_{out}) = net\ heat\ in\ [J]$
$\Sigma W_{out} = (W_{out} - W_{in}) = net\ work\ out\ [J]$
$\Delta U = change\ in\ total\ energy\ [J]$
$\Delta U = change\ in\ internal\ energy\ [J]$
$\Delta KE = change\ in\ kinetic\ energy\ [J]$
$\Delta PE = change\ in\ potential\ energy\ [J]$

The **first law** of thermodynamics describes *conservation of energy*. The total energy of an isolated system is constant; energy can be transformed from one form to another, but can never be created nor destroyed.

- It assumes that the total energy input to the system is equal to the total energy output. The expression above is not applicable to flows with internal release of heat within the fluid due to chemical or nuclear reactions.

$Q = mc_v(T_2 - T_1)$ CONSTANT VOLUME PROCESS

$Q = h_2 - h_1 = mc_p(T_2 - T_1)$ CONSTANT PRESSURE PROCESS

$pv^n = constant$ POLYTROPIC PROCESS

$W = \dfrac{p_1 v_1 - p_2 v_2}{n - 1}$ WORK DONE (POLYTROPIC)

Internal Energy

$$\Delta U = mc\Delta T$$

$m = mass\ [kg]$
$c = specific\ heat\ capacity\ [J\ kg^{-1}\ K^{-1}]$
$T = temperature\ [K]$

For processes which occur at constant pressure, $c = c_p$.
For water, $c_p = 4185.5\ J\ kg^{-1}\ K^{-1}$.

7.2. SECOND LAW

$$\Delta S = \frac{Q}{T} \geq 0$$

The **second law** of thermodynamics states that *the total entropy of an isolated system can never decrease over time.*

Entropy change for an adiabatic reversible process

$$\Delta S = 0$$

Entropy change for an irreversible process

$$\Delta S_{gen} = \Delta S_{total} = \Delta S_{sys} + \Delta S_{surr} \geq 0$$

Real world processes generate entropy, in the system or in the surroundings.

Heat transferred across a thermal reservoir (T=constant)

$$Q = T\Delta S$$

Heat generated throughout a reversible process

$$Q_{reversible} = \int T ds$$

All complex natural processes are irreversible. These irreversibilities (i.e. losses) include all heat transfer through a finite (non-zero) temperature difference, friction, plastic deformation, flow of electric current through a resistance, magnetization or polarization with a hysteresis, unrestrained expansion of fluids, spontaneous chemical reactions, or spontaneous mixing of matter of varying composition/states.

Clausius inequality for a cyclic process

$$\oint \frac{\delta Q}{T} \leq 0$$

No system can produce a net amount of work while operating in a cycle and exchanging heat with a single thermal reservoir (an engine will always need the equivalent of a **heat source** and a **heat sink**).

For a reversible cyclic process, there is no generation of entropy in each of the infinitesimal heat transfer processes (the equality holds). However, for all irreversible cyclic processes (real-world processes), net entropy is always generated.

7.3. IDEAL GAS EQUATIONS

Ideal Gas Law

$$pV = mRT \quad \text{or} \quad pv = RT$$

$p = pressure\ [Pa]$
$V = volume\ [m^3]$
$m = mass\ [kg]$
$R = individual\ gas\ constant\ [J\ kg^{-1}\ K^{-1}]$
$T = temperature\ [K]$
$v = specific\ volume = V/m = 1/\rho\ [m^3\ kg^{-1}]$

Combined Gas Law

$$\frac{p_1 V_1}{T_1} = \frac{p_2 V_2}{T_2}$$

$$R = \frac{R_0}{M}$$

INDIVIDUAL GAS CONSTANT

$R_0 = universal\ gas\ constant\ [8.314\ kJ\ kmol^{-1}\ K^{-1}]$
$M = molar\ mass\ [kg\ mol^{-1}]$

Individual gas constants and specific heats (at 1 atm, 300K)

Gas		R $[J\ kg^{-1}\ K^{-1}]$	c_p $[kJ\ kg^{-1}\ K^{-1}]$	c_v $[kJ\ kg^{-1}\ K^{-1}]$
Air	-	**287.0**	**1.005**	**0.718**
Argon	Ar	208.1	0.520	0.312
Ammonia	NH_3	488	2.164	1.645
Butane	C_4H_{10}	144.3	1.716	1.573
Carbon Dioxide	CO_2	188.9	0.846	0.657
Carbon Monoxide	CO	296.8	1.040	0.744
Helium	He	2 076.9	5.193	3.116
Hydrogen	H_2	4 124.0	14.307	10.183
Methane	CH_4	518.2	2.254	1.735
Nitrogen	N_2	296.8	1.039	0.743
Oxygen	O_2	259.8	0.918	0.658
Propane	C_3H_8	188.5	1.679	1.491
Water Vapour	H_2O	461.5	1.872	1.411

Specific Heat Capacities

$$\left(\frac{\partial h}{\partial T}\right)_{p\,=\,const} = c_p,\ \left(\frac{\partial u}{\partial T}\right)_{v\,=\,const} = c_v$$

$$c_p - c_v = R$$

$$\gamma = \frac{c_p}{c_v}$$

$$u = c_v T$$

$$h = c_p T$$

For air, $\gamma = 1.4$ up to around $811\ K$ ($1{,}000\ ^\circ F$ or $538\ ^\circ C$).

7.4. ISENTROPIC PROCESSES IN GAS TURBINES

$$pv^{\gamma} = constant$$

$$Tv^{\gamma - 1} = constant$$

$$\frac{T_1}{T_2} = \left(\frac{p_1}{p_2}\right)^{\frac{\gamma - 1}{\gamma}};\quad \left(\frac{p_1}{p_2}\right) = \left(\frac{\rho_1}{\rho_2}\right)^{\gamma}$$

7.5. CYCLE EFFICIENCY

$$\eta = \frac{W_{net}}{Q_{in}} = 1 - \frac{Q_{out}}{Q_{in}}$$

$$\eta_{real} = 1 - \frac{Q_C}{Q_H}$$

Ideal Cycle Efficiency

$$\eta_{Otto} = 1 - \frac{1}{r_v^{\gamma - 1}}$$

$$\eta_{Brayton} = 1 - \left(\frac{1}{r_p}\right)^{\frac{\gamma - 1}{\gamma}}$$

$$\eta_{diesel} = 1 - \frac{1}{r_v^{\gamma - 1}} \frac{r_c^{\gamma} - 1}{\gamma(r_c - 1)}$$

7.6. HEAT PUMPS AND REFRIGERATION

Coefficient of Performance

$$COP_{ref, real} = -\frac{Q_{in}}{W}(+ve) = \frac{Q_C}{Q_H - Q_C}$$

$$COP_{hp, real} = \frac{Q_{out}}{W}(+ve) = \frac{Q_C}{Q_H - Q_C}$$

7.7. CARNOT CYCLE

$$\eta_{ideal} = \eta_{Carnot} = 1 - \frac{T_{cold}}{T_{hot}}$$

$$COP_{hp, ideal} = COP_{Carnot\ heat\ pump} = \frac{T_{hot}}{T_{hot} - T_{cold}}$$

$$COP_{ref, ideal} = COP_{Carnot\ refrigerator} = \frac{T_{cold}}{T_{hot} - T_{cold}}$$

7.8. HEAT TRANSFER

Sensible Heat

$$Q = mc\Delta T$$

$Q = heat\ transfer\ [W]$
$c = specific\ heat\ capacity\ [J\ kg^{-1}\ K^{-1}]$

Conduction

Fourier rate law for conduction

$$\dot{Q} = -kA\frac{dT}{dx}$$

$$\dot{Q} = -\frac{kA\Delta T}{l}$$

$k = heat\ transfer\ coefficient\ [W\ m^{-1}\ K^{-1}]$
$A = heat\ transfer\ area\ [m^2]$
$T = temperature\ [K]$
$x = distance\ along\ conductor\ in\ direction\ of\ heat\ transfer\ [m]$
$l = total\ length\ [m]$

Convection

$$\dot{Q} = hA\Delta T$$

$\dot{Q} = heat\ flow\ [W]$
$h = heat\ transfer\ coefficient\ [W\ m^2\ K^{-1}]$
$A = heat\ transfer\ area\ [m^2]$
$\Delta T = temperature\ difference\ [K]$

Composite Slab or Laminate with Fluid Boundaries

$$\dot{Q} = \frac{A\Delta T}{\frac{1}{h_{fa}} + \frac{d_1}{k_1} + \frac{d_2}{k_2} + \frac{d_3}{k_3} + \ldots + \frac{1}{h_{fb}}}$$

$d = slab\ width\ [m]$
$h = convection\ coefficient$

Solid Expansion

$$\Delta l = \alpha l_0 \Delta T$$

$\alpha = thermal\ expansion\ coefficient\ [m \cdot m^{-1} K^{-1}]$
$l_0 = original\ length$

Work

Linear work

$$W = \int_{r1}^{r2} \vec{F} \cdot d\vec{s}, \quad W = \overline{F} \Delta s \cos\theta$$

$\vec{F} = net\ force\ [N]$
$\vec{s} = displacement\ [m]$

Rotary work

$$W = \int_{\theta 1}^{\theta 2} \vec{\tau} \cdot d\vec{\theta}$$

$\vec{\tau} = torque\ [N\ m]$
$\vec{\theta} = angular\ displacement\ [rad]$

Power

$$\overline{P} = \frac{\Delta W}{\Delta t}, \quad P = \frac{dW}{dt}$$

$\overline{P} = average\ power\ [kW]$
$P = instantaneous\ power\ [kW]$
$W = work\ [kJ]$
$t = time\ [s]$

7.9. FLOW EQUATIONS

Change in Enthalpy

Incompressible liquids with constant specific heat

$$h_2 - h_1 = c(T_2 - T_1) + v(p_2 - p_1)$$

$c = specific\ heat\ [J\ kg^{-1}\ K^{-1}]$
$v = volume\ [m^3\ kg^{-1}]$
$p = pressure\ [Pa]$

Steady Flow Energy Equation (open system)

$$\Delta Q - \Delta W = \dot{m}\left((h_2 - h_1) + \frac{1}{2}\left(v_2{}^2 - v_1{}^2\right) + (gz_2 - gz_1)\right)$$

$\Delta Q = net\ heat\ flow\ in\ [W] = (Q_{in} - Q_{out})$
$\Delta W = net\ work\ out\ [W]\ (W_{out} - W_{in})$
$\dot{m} = mass\ flow\ rate\ [kg\ s^{-1}]$
$h = enthalpy\ [J]$
$v = velocity\ [m\ s^{-1}]$
$g = gravitational\ constant\ (9.08665\ m\ s^2)$
$z = height\ [m]$

The **Steady Flow Energy Equation (SFEE)** is a consequence of the **First Law**, expressing the total energy flow rate into the system (i.e. heat input minus work output) for an open system control volume.

- It is assumed that the mass flow through the system is constant.
- It is also assumed that the total energy input to the system is equal to the total energy output. It is not applicable to flows with internal release of heat within the fluid due to chemical or nuclear reactions.

7.10. DIFFUSION

Fick's First Law of Diffusion

$$J = -D\frac{d\varphi}{dx}$$

$J = diffusion\ flux\ [mol\ m^{-2} s^{-1}]$
$D = diffusivity\ (diffusion\ coefficient)\ [m^2\ s^{-1}]$
$\varphi = concentration\ [mol\ m^{-3}]$
$x = position\ [m]$

Fick's Second Law of Diffusion

$$\frac{d\varphi}{dt} = D\frac{\partial^2 \varphi}{\partial x^2}$$

$t = time\ [s]$

The Arrhenius Equation

$$D = D_0 e^{-\left(\frac{E_A}{RT}\right)}$$

$D_0 = maximal\ diffusion\ coefficent\ (at\ infinite\ temperature)\ [m^2 s^{-1}]$
$E_A = activation\ energy\ for\ diffusion\ [J\ mol^{-1}]$

> To increase a rate of diffusion, either increase temperature, or decrease the activation energy (e.g. via catalyst).

Alternatively, the general Arrhenius Equation:

$$k = Ae^{-\left(\frac{E_a}{RT}\right)}$$

$k = rate\ constant$
$T = absolute\ temperature\ [K]$
$A = pre-exponential\ factor$
$E_a = activation\ energy\ (in\ same\ units\ as\ RT)$
$R = universal\ gas\ constant\ (\sim 8.31446\ J\ K^{-1}\ mol^{-1})$

> A rough rule of thumb is that corrosion rate doubles for each 10 °C rise in temperature (Shifler and Aylor, Considerations for the Testing of Materials and Components in Seawater, Corrosion/2002, Paper 217, Houston, TX, 2002).

8. FLUID MECHANICS

8.1. STATICS

Static Pressure

The change in static pressure in a barotropic, compressible fluid is given by:

$$\Delta p = - g \int_{h_1}^{h_2} \rho dh$$

For an incompressible fluid, this simplifies to:

$$\Delta p = - \rho g \Delta h$$

$\rho = density\ of\ the\ fluid\ [kg\ m^{-3}]$
$g = acceleration\ due\ to\ gravity\ (9.80665\ m\ s^{-2})$
$h = vertical\ height\ above\ an\ arbitrary\ datum\ [m]$

> A barotropic fluid is one whose pressure and density are related by an equation of state that does not contain temperature as a dependent variable.

The static pressure p_2 in the manometer below is given by:

$$p_2 = p_0 + \rho_1 g \Delta h_1 + \rho_2 g \Delta h_2$$

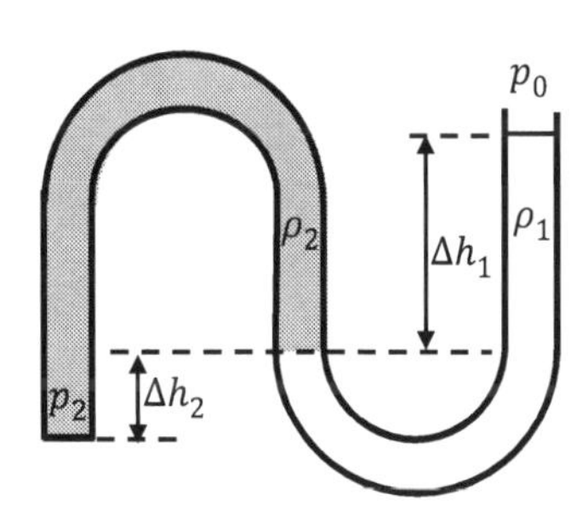

$\rho = density\ [kg\ m^{-3}]$
$g = acc'n\ due\ to\ gravity\ (9.80665\ m\ s^{-2})$

Buoyancy Force

$$F_B = \rho g V$$

$V = volume\ of\ displaced\ fluid\ [m^2]$

Surface Tension

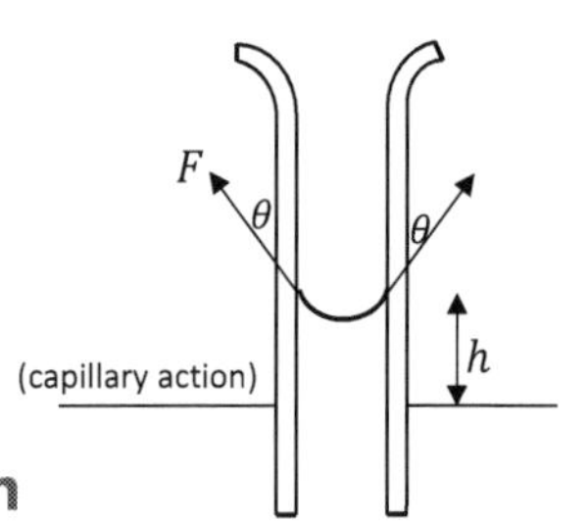

$$2\pi r\gamma = F\cos\theta \approx \pi r^2\rho g h$$

$\gamma = surface\ tension\ [N\,m^{-1}]$

Needle supported by surface tension

$$W = 2\gamma L\cos\theta$$

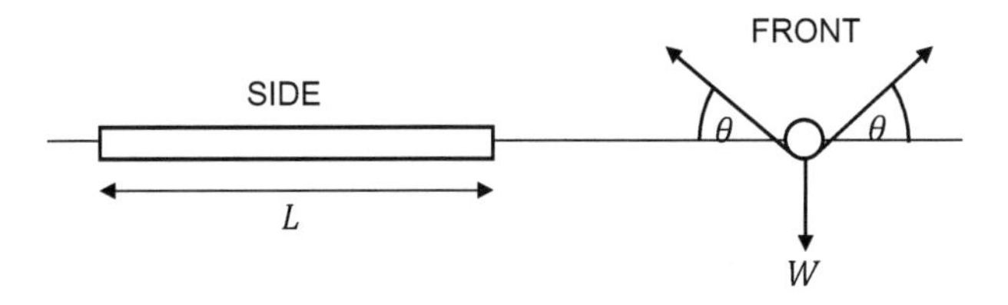

Pressure of a drop or bubble

Liquid Drop

$$P_i - P_o = \frac{2\gamma}{R}$$

Hollow Bubble

$$P_i - P_o = \frac{4\gamma}{R}$$

(Two surfaces)

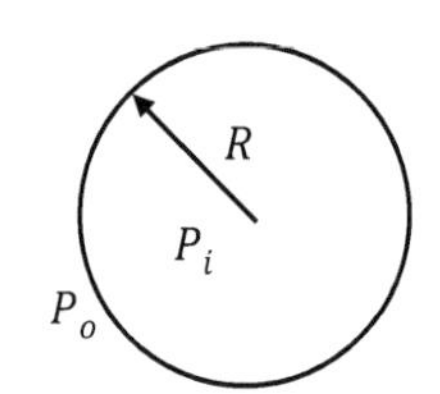

Fluid		**Surface tension** $[10^{-3}\ N\,m^{-1}]$
Acetone (propanone)	C_3H_6O	23
Crude oil, light	Hydrocarbons	32
Crude oil, heavy	Hydrocarbons	37
Ethanol (Ethyl alcohol)	C_2H_6O	22.3
Ethylene glycol	$C_2H_6O_2$	48.8
Mercury	Hg	465
Water	H_2O	72.8

Properties are at 20°C. For comparison, water at 100°C has a surface tension of $59 \times 10^{-3}\ N\,m^{-1}$.

8.2. DYNAMICS

Stagnation Pressure

$$p_{stagnation} = p_{static} + p_{dynamic} = p + \frac{1}{2}\rho v^2$$

$v = velocity\ [m\ s^{-1}]$

Bernoulli's Equation

$$p + \frac{1}{2}\rho v^2 + \rho g h = constant$$

For inviscid, incompressible, steady, irrotational flow, the sum of stagnation pressure and elevation pressure is constant along a streamline.

$$H_{total} = \frac{p}{\rho g} + \frac{v^2}{2g} + h$$

The total head (i.e. internal energy) of a fluid is comprised of the static pressure head, velocity head, and elevation head.

Mass Continuity

$$\dot{m}_1 = \dot{m}_2$$

$$\rho_1 Q_1 = \rho_2 Q_2$$

$$\rho_1 A_1(\vec{v}_1 \cdot \hat{n}_1) = \rho_2 A_2(\vec{v}_2 \cdot \hat{n}_2)$$

$Q = volumetric\ flow\ rate\ [m^3\ s^{-1}]$
$A = surface\ area\ [m^2]$
$\vec{v} = velocity\ [m\ s^{-1}]$
$\vec{n} = unit\ vector\ normal\ to\ the\ entrance\ or\ exit\ area$

Dynamic Viscosity

For an isotropic Newtonian fluid

$$\tau = \mu \frac{du}{dy}$$

$\tau = shear\ stress\ [Pa]$
$\mu = dynamic\ viscosity\ [Pl] = [kg\ m^{-1}\ s^{-1}]$
$u = tangential\ velocity\ [m\ s^{-1}]$
$y = distance\ from\ wall\ [m]$

Reynolds Number

$$Re = \frac{\rho v D}{\mu}$$

$\rho = density$
$v = characteristic\ (free\ stream)\ velocity$
$D = characteristic\ length$

> The Reynold's number can be described as a ratio of inertial forces to viscous forces. Above the critical number, flow becomes fully turbulent. For pipes, use $Re_{crit} \approx 2300$.

Pipe Friction - D'Arcy's Formula

Head loss through a pipe

$$h_f = f\frac{Lv^2}{2gD}$$

Pressure loss through a pipe

$$\Delta p = f\frac{L}{D}\frac{\rho v^2}{2}$$

$f = Darcy - Weisbach\ friction\ factor\ (see\ Moody\ diagram).$

Head loss through a fitting

$$h_L = K\frac{v^2}{2g}$$

$K = K\ factor\ (loss\ coefficient),\ specific\ to\ each\ type\ of\ valve$

Laminar Friction Factor

The Darcy friction factor for laminar flow in circular pipes is given by:

$$f = \frac{64}{Re}$$

> For laminar flow, the head loss is proportional to velocity rather than velocity squared, thus the friction factor is inversely proportional to velocity (or Reynold's number).

Stokes Drag

Flow past a sphere at $Re < 2$

$$F_{drag} = 3\pi D v_{\infty}\mu$$

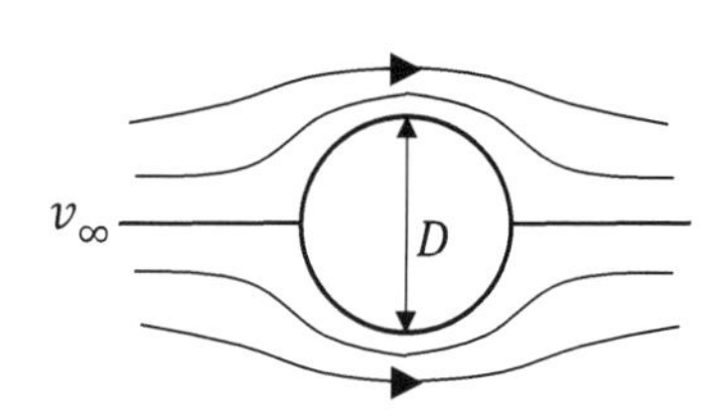

Pipe Roughness

Pipes (Material)	**Absolute Roughness microns** $[10^{-6} m]$
Drawn tubing (glass, brass, plastic)	1.5
Copper	1.5
Aluminium	1.5
PVC	1.5
Red brass	1.5
Fiberglass	5
Carbon steel or wrought iron	45
Stainless steel	45
Cast iron-asphalt dipped	120
Galvanized iron	150
Cast iron uncoated	250
Wood stave	100-200
Ductile iron	2,500
Concrete	300 - 3,000
Riveted steel	1,000 - 10,000

Fittings	L/D
Globe valve	340
Gate valve	8
Lift check valve	600
Swing check valve	50 - 100
Ball valve	6
Butterfly valve	35
Flush pipe entrance (sharp corner)	K=0.5
Flush pipe entrance (radius >0.15)	K=0.04
Pipe exit	K=1
Tee through	20
Tee branch flow	60
Elbow (90 degrees)	30
Elbow (45 degrees)	16

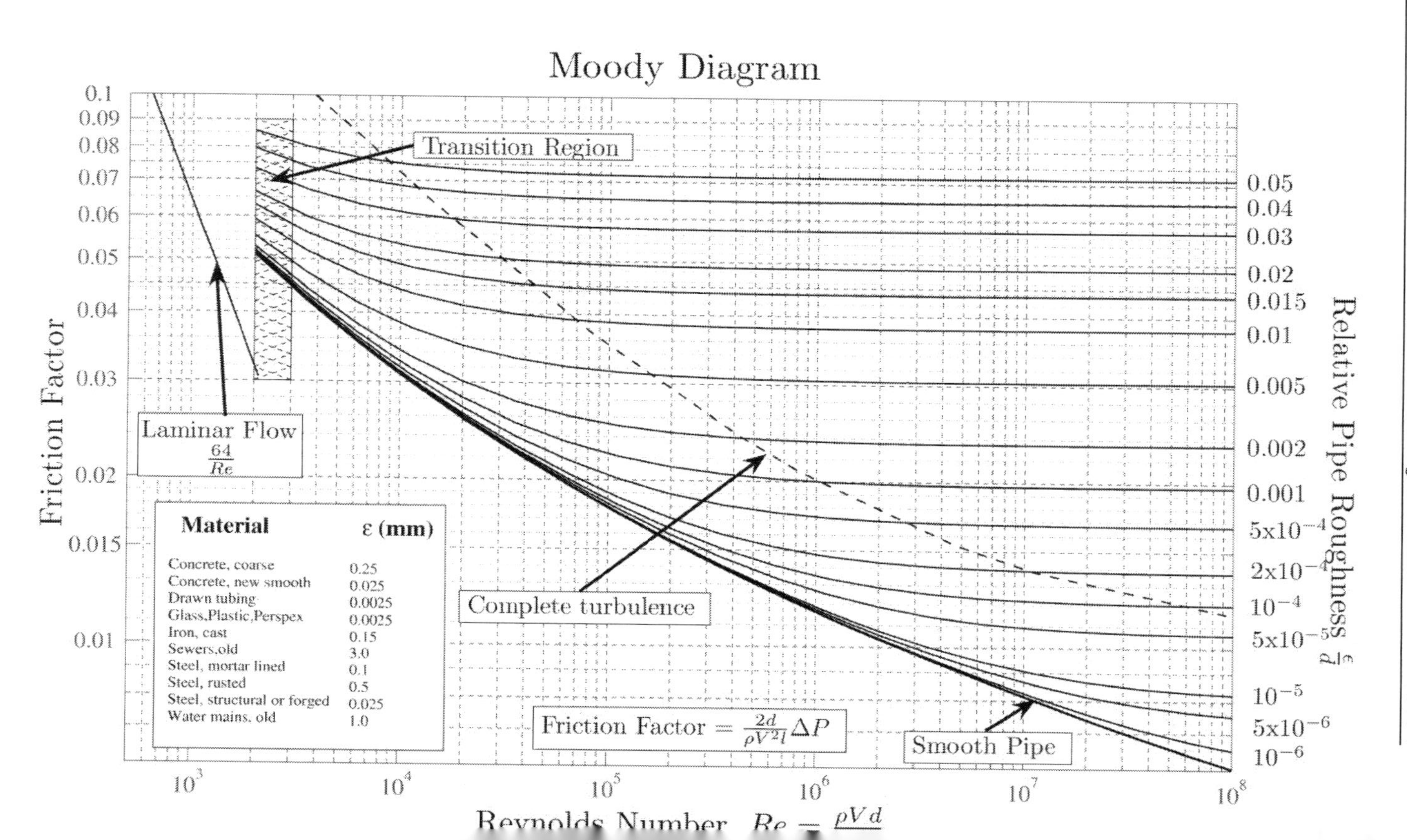
Moody Diagram
Relative Pipe Roughness
0.05
0.04
0.03
0.02
0.015
0.01
0.005
0.002
0.001
5x10⁻⁴
2x10⁻⁴
10⁻⁴
5x10⁻⁵
10⁻⁵
5x10⁻⁶
10⁻⁶
Transition Region
Laminar Flow 64/Re
Complete turbulence
Smooth Pipe
Friction Factor = 2d/(ρV²l) ΔP
Friction Factor
0.1
0.09
0.08
0.07
0.06
0.05
0.04
0.03
0.02
0.015
0.01
10³
10⁴
10⁵
10⁶
10⁷
10⁸
Material
ε (mm)
Concrete, coarse 0.25
Concrete, new smooth 0.025
Drawn tubing 0.0025
Glass,Plastic,Perspex 0.0025
Iron, cast 0.15
Sewers,old 3.0
Steel, mortar lined 0.1
Steel, rusted 0.5
Steel, structural or forged 0.025
Water mains, old 1.0

K-Factor of a Pipe Fitting

Fitting	Types	K
45° Elbow	Standard (R/D = 1)	0.4
	Long Radius (R/D = 1.5)	0.2
90° Elbow Curved	Standard (R/D = 1)	0.8
	Long Radius (R/D = 1.5)	0.5
90° Elbow Square or Mitred		1.3
180° Bend	Close Return	1.5
Tee, Run Through	Branch Blanked	0.4
Tee, as Elbow	Entering in run	1
Tee, as Elbow	Entering in branch	1
Tee, Branching Flow		1
Gate valve	Fully Open	0.2
	3/4 Open	0.9
	1/2 Open	4.5
	1/4 Open	24
Diaphragm valve	Fully Open	2.3
	3/4 Open	2.6
	1/2 Open	4.3
	1/4 Open	21
Globe valve, Bevel Seat	Fully Open	6
	1/2 Open	9.5
Plug valve	$\theta = 5°$	0.1
	$\theta = 10°$	0.3
	$\theta = 20°$	1.6
	$\theta = 40°$	17
	$\theta = 60°$	206
Butterfly valve	$\theta = 5°$	0.2
	$\theta = 10°$	0.5
	$\theta = 20°$	1.5
	$\theta = 40°$	11
	$\theta = 60°$	118
Check valve	Swing	2
	Disk	10
	Ball	70

Mass Continuity (Control Volume)

Integral Form

$$\frac{dm}{dt} = \frac{\partial}{\partial t}\int_{CV} \rho dV + \int_{A} \rho(\vec{v}\cdot\vec{n})dA = 0$$

$V = volume$
$\vec{v} = velocity$
$\vec{n} = normal\ vector$
$A = surface\ area$

> The mass change within a control volume plus the mass flowing out of the surface of the control volume equals zero.

For incompressible flows, there is no *mass change* term:

$$\frac{dm}{dt} = \int_{A} \rho(\vec{v}\cdot\vec{n})dA = 0$$

Differential Form

$$\frac{\partial \rho}{\partial t} + \nabla\cdot(\rho u) = 0$$

For incompressible flows, the divergence equals zero:

$$\nabla\cdot u = 0$$

Momentum Continuity (Control Volume)

$$\frac{d(m\vec{v})}{dt} = \frac{\partial}{\partial t}\int_{CV} \rho\vec{v} dV + \int_{A} \rho\vec{v}(\vec{v}\cdot\vec{n})dA = 0$$

Navier-Stokes Equation

$$\rho\left(\frac{\partial(u)}{\partial t} + u\cdot\nabla u\right) = -\nabla p + \mu\nabla^2 u + \rho g$$

> The Navier-Stokes Equations are analogous to Newton's Second Law applied to an infinitesimal unit volume. The LHS includes the **material derivative** and is analogous to *mass times acceleration*. The RHS is analogous to *net force*.

> The material derivative of an infinitesimal fluid packet is equal to the sum of the negative pressure gradient, shear forces, and body forces.

8.3. AERODYNAMICS

Lift and Drag Coefficient

$$C_L = \frac{F_L}{\frac{1}{2}\rho v_\infty{}^2 A} = \frac{F_L}{qA} \qquad C_D = \frac{F_D}{\frac{1}{2}\rho v_\infty{}^2 A} = \frac{F_D}{qA}$$

$q = dynamic\ pressure = \frac{1}{2}\rho v^2$

$A = plan\ area\ [m^2]$

Induced Drag Coefficient

$$C_{Di} = \frac{C_L^2}{\pi e AR}$$

$AR = aspect\ ratio$
$e = wing\ span\ efficiency$

e is the wing span efficiency value by which the induced drag exceeds that of an elliptical lift distribution, typically 0.95-0.99.

Speed of Sound

$$c = \sqrt{\gamma RT}$$

γ *is the ratio of specific heats* (1.4 *for air*)
R *is the ideal gas constant* $(286.9\ J\ kg^{-1}\ K^{-1}\ for\ air)$
T *is the local air temperature* $[K]$

Mach Number

$$M = \frac{v}{c}$$

$v = velocity\ [m\ s^{-1}]$
$c = speed\ of\ sound\ [m\ s^{-1}]$

Air is generally considered incompressible below $M = 0.3$.

Ram Air Recovery

To calculate ideal ram air temperature and pressure recovery (e.g. NACA scoops, engine intakes):

$$T_T = T_0\left(1 + \left(\frac{\gamma - 1}{2}\right)M^2\right)$$

$$P_T = P_0\left(1 + \left(\frac{\gamma - 1}{2}\right)M^2\right)^{\frac{\gamma}{\gamma - 1}}$$

$T_T = total\ temperature\ [K]$
$and\ P_T\ are\ total\ temperature\ and\ pressure\ respectively$
$T_0\ and\ P_0\ are\ free\ stream\ temperature\ and\ static\ pressure\ respectively$
$\gamma\ is\ the\ ratio\ of\ specific\ heats\ (1.4\ for\ air)$

The actual recovery pressure is slightly less than the total pressure due to losses:

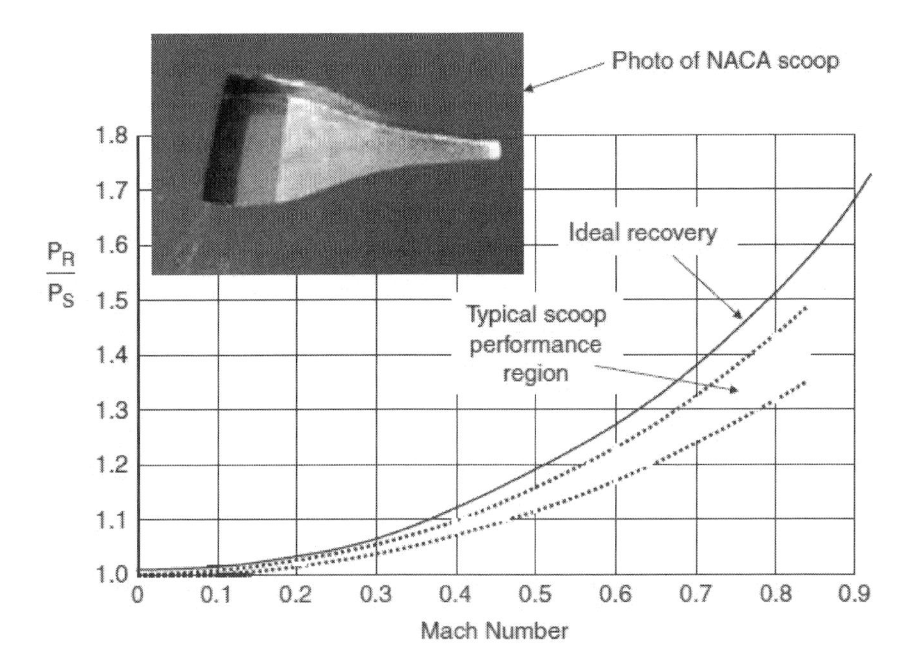

8.4. JET ENGINES

Propulsive Power

$$P = T \cdot V_\infty = \dot{m}(V_e - V_\infty) \cdot V_\infty$$

$T = thrust\ [N]$
$\dot{m} = mass\ flow\ rate\ of\ air\ [kg\ s^{-1}]$
$V_\infty = free\ stream\ velocity\ [m\ s^{-1}]$
$V_e = average\ exhaust\ velocity\ [m\ s^{-1}]$

Thrust Specific Fuel Consumption

$$sfc = \frac{\dot{m}_f}{T}$$

$\dot{m}_f = fuel\ flow\ rate\ [kg\ s^{-1}]$

Propulsive Efficiency

$$\eta_p = \frac{2}{1 + \frac{v_e}{v_\infty}}$$

Thermal Efficiency

$$\eta_t = 1 - \left(\frac{p_1}{p_2}\right)^{\frac{\gamma - 1}{\gamma}} = 1 - (r_p)^{\frac{1-\gamma}{\gamma}}$$

$p_1 = compressor\ inlet\ pressure\ [kPa]$
$p_2 = combustor\ inlet\ pressure\ [kPa]$
$r_p = compressor\ pressure\ ratio$

Overall Efficiency

$$\eta_o = \eta_p \eta_t$$

9. SYSTEMS

Transfer Function

$$G(s) = \frac{Y(s)}{X(s)} = \frac{L\{y(t)\}}{L\{x(t)\}}$$

First Order Systems

$$\tau \frac{dy}{dt} + y(t) = x(t)$$

$t = time\ [s]$
$\tau = time\ constant$
$f(t) = forcing\ function$

Response to a Step Input

Boundary conditions:

$$y(0) = 0$$

Excitation:

$$x(t) = \begin{cases} 0\ for\ t < 0 \\ H\ for\ t \geq 0 \end{cases}$$

Solution for $y(0)=0$, for step input $x(t \geq 0) = H$, is

$$y(t) = H(1 - e^{-t/\tau})$$

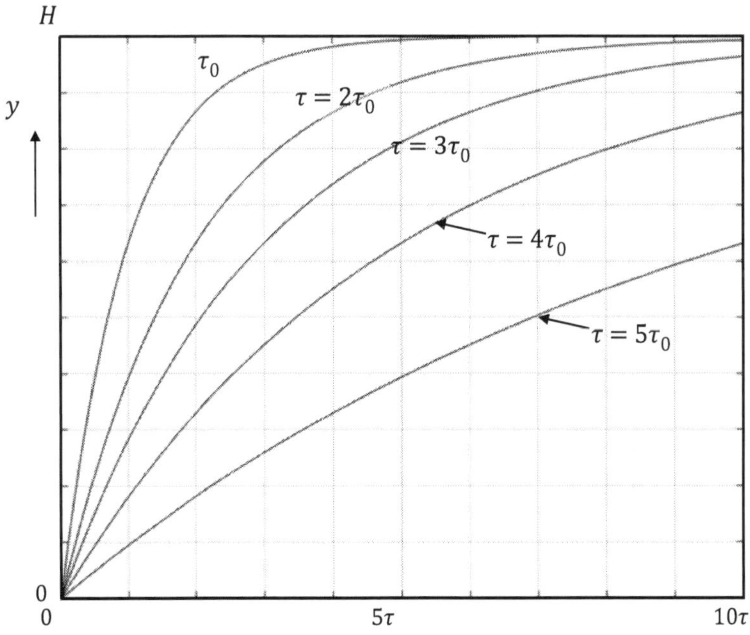

Second Order Systems

$$\frac{1}{\omega_n^2}\frac{d^2y}{dx^2} + \frac{2\zeta}{\omega_n}\frac{dy}{dx} + y(t) = x(t)$$

$\omega_n = undamped\ natural\ frequency\ [rad\ s^{-1}]$
$\zeta = damping\ ratio\ (i.e.\ damping\ factor)$
$x(t) = excitation\ (input)$
$y(t) = system\ response\ (output)$

$$\zeta = \frac{c}{2m\omega_n}$$

Undamped: (ζ=0)

Underdamped: (ζ<1)

Critically damped (ζ=1),

Overdamped (ζ>1).

Response to a Step Input

Boundary conditions:

$$y(0) = \frac{dy(0)}{dt} = 0$$

Excitation:

$$x(t) = \begin{cases} 0\ for\ t < 0 \\ H\ for\ t \geq 0 \end{cases}$$

$\omega_d = \omega_n\sqrt{1 - \zeta^2}$ DAMPED NATURAL FREQUENCY

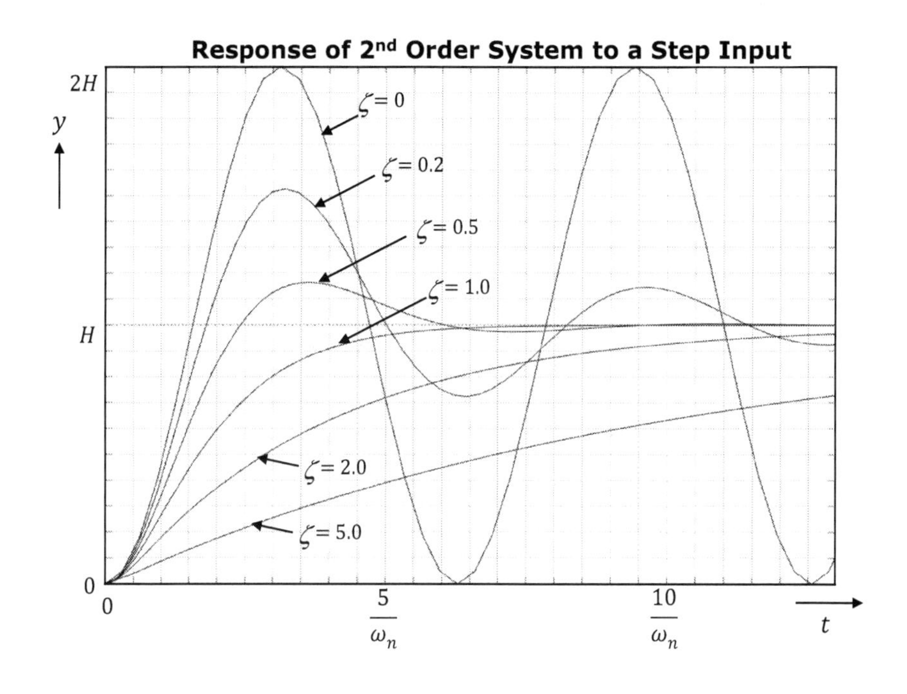

Frequency Response (Gain)

$$G(j\omega) = \frac{1}{\left(1 + \left(\frac{\omega}{\omega_n}\right)^2\right) + 2\zeta\frac{j\omega}{\omega_n}}$$

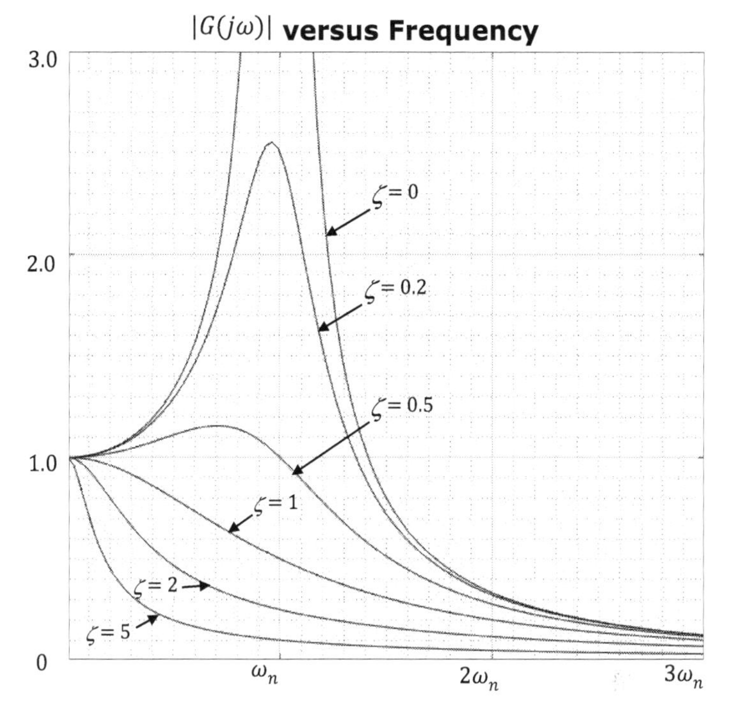

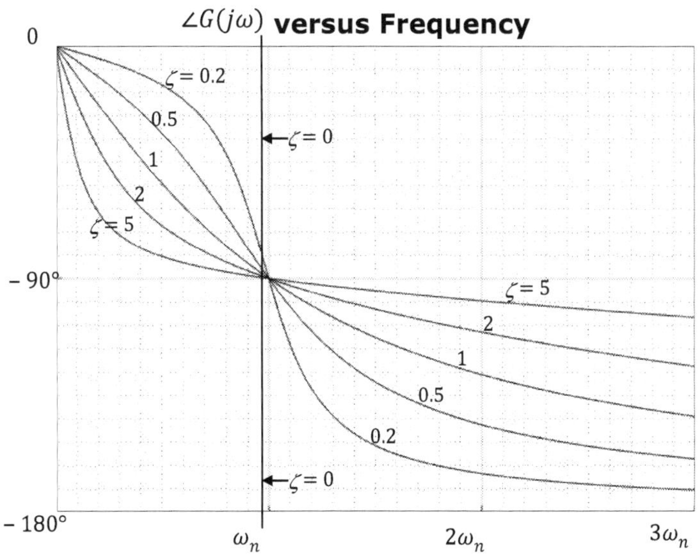

10. PROJECT MANAGEMENT

Earned Value Analysis

$PV = (Planned\ \%\ complete) \times BAC$

The Planned Value is the planned percent already complete multiplied by the Budget At Completion. It may also be referred to as *'Budgeted Cost of Work Scheduled'*. Alternatively, it is the 'time-phased baseline of the value of the work scheduled'. Planned Value is an approved cost estimate of the resources scheduled to be used during a project.

Budget at Completion (BAC) is the total budget allocated to the project.

$EV = (\%\ completed) \times BAC$ EARNED VALUE

Earned Value for a task is simply the percent complete times its original budget. Stated differently, EV is the percent of the original budget that has been earned by actual work completed. It may also be referred to as *'Budgeted Cost of Work Performed'* and gives some indication of how much value you have actually delivered so far.

$CV = EV - AC$ COST VARIANCE

Cost variance indicates if the work accomplished using labour and materials costs more or less than was planned at any point in the project. You may decide to track the Cost Variance over a project.

$SV = EV - PV$ SCHEDULE VARIANCE

Schedule Variance presents an overall assessment in monetary terms of the progress of all work packages in the project. Schedule variance is the difference between the earned value and the baseline planned value to date. You may decide to track the Schedule Variance over a project.

Performance Indices

$$CPI = \frac{EV}{AC}$$

COST PERFORMANCE INDEX

Cost Performance Index measures the cost efficiency of work accomplished to date (Earned Value/Actual Costs). If you track CPI during a project, you may use it to decide when to implement cost control measures, for example if it starts to deviate too far from 1.00 due to special causes.

$$SPI = \frac{EV}{PV}$$

SCHEDULE PERFORMANCE INDEX

Schedule Performance Index measures scheduling efficiency (Earned Value/Planned Value). If you track SPI during a project, you may use it to decide when to implement schedule control measures, for example if it starts to deviate too far from 1.00 due to special causes.

Index	Cost (CPI)	Schedule (SPI)
>1.00	Under cost	Ahead of schedule
=1.00	On cost	On schedule
<1.00	Over cost	Behind schedule

$$EAC = \frac{BAC}{CPI}$$

ESTIMATE AT COMPLETION

The estimated cost at completion extrapolates based on what your budget was and how cost-efficient you have been to date.

$EV = earned\ value\ (BCWP\ Budgeted\ Cost\ of\ Work\ Performed)$
$AC = actual\ costs\ (ACWP\ Actual\ Cost\ of\ Work\ Performed)$
$PV = planned\ value\ (BCWS\ Budgeted\ Cost\ of\ Work\ Scheduled)$
$BAC = budget\ at\ completion$

Percent Complete Indices

$$PCIB = \frac{EV}{BAC}$$

%COMPLETE INDEX - BUDGET

The PCIB is the percentage of work complete compared to Baseline budget.

$$PCIC = \frac{AC}{BAC}$$

%COMPLETE INDEX - COST

The PCIC is the percentage of work complete as actual costs compared with the revised estimate of project costs.

$$Estimated\ Time\ to\ Complete = \frac{Original\ Time\ Estimate}{SPI}$$

$EV = earned\ value\ (BCWP\ Budgeted\ Cost\ of\ Work\ Performed)$
$BAC = budget\ at\ completion$
$AC = actual\ costs\ (ACWP\ Actual\ Cost\ of\ Work\ Performed)$
$SPI = schedule\ performance\ index$

In the chart below, this particular project has a negative cost and schedule variance, on track to delivering late and over-budget.

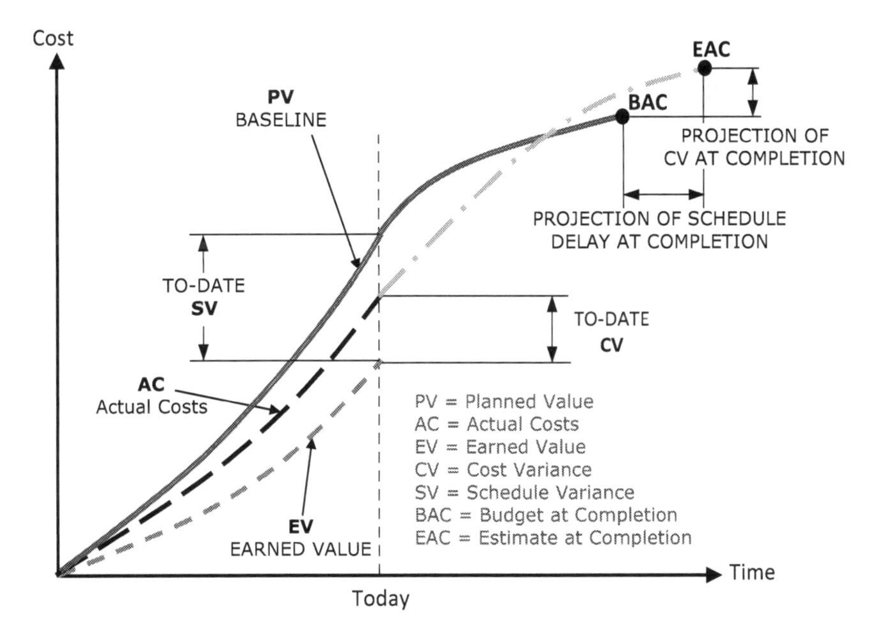

PROPERTIES OF WATER

T	Density	C_p	Vapour Pressure	Dynamic Viscosity	Thermal Conduct.	Surface Tension
$°C$	$kg\,m^{-3}$	$J\,kg^{-1}\,K^{-1}$	kPa	$10^{-6}\,Pa\,s$	$mW\,K^{-1}\,m^{-1}$	$mN\,m^{-1}$
0	999.84	4217.6	0.6113	1793	561.0	75.64
10	999.70	4192.1	1.2281	1307	580.0	74.23
20	998.21	4181.8	2.3388	1002	598.4	72.75
30	995.65	4178.4	4.2245	797.7	615.4	71.20
40	992.22	4178.5	7.3814	653.2	630.5	67.60
50	988.03	4180.6	12.344	547.0	643.5	67.94
60	983.20	4184.3	19.932	466.5	654.3	66.24
70	977.78	4189.5	31.176	404.0	663.1	64.47
80	971.82	4196.3	47.373	354.4	670.0	62.67
90	965.35	4205.0	70.177	314.5	675.3	60.82
100	958.40	4215.9	101.325	281.7	679.1	58.91

ATMOSPHERE

International Standard Atmosphere (ISA)

Mean Sea Level Conditions

Symbol	Value	Description
P_0	101 325 *Pa*	Pressure
ρ_0	1.225 $kg\ m^{-3}$	Density
T_0	288.15 *K*	Temperature
a_0	340.294 $m\ s^{-1}$	Speed of sound
g_0	9.80665 $m\ s^{-2}$	Standard gravity

Troposphere Model

$Up\ to\ 11{,}000\ m\ (36{,}089\ ft)$

Temperature

$$T = T_0 - 6.5\frac{h}{1{,}000}$$

$T = air\ temperature\ [K]$
$h = altitude\ above\ sea\ level\ [m]$

Pressure

$$P = P_0\left(1 - 0.0065\frac{h}{T_0}\right)^{5.2561}$$

Density

$$\rho = \frac{p}{R_s T}$$

$p = air\ pressure\ [kPa]$
$R_s = specific\ gas\ constant = 287.058\ J\ kg^{-1}\ K^{-1}\ for\ dry\ air$

$T = air\ temperature\ [K]$

Standard Atmosphere Table

Elevation	Temperature	Pressure	Relative Density	Kinematic Viscosity	Thermal Conductivity	Speed of Sound
[m]	[K]	[bar]	[1]	[10^{-5} m^2 s^{-1}]	[10^{-2} $Wm^{-1}K^{-1}$]	[m/s]
h	T	P	ρ/ρ_o	v	κ	a
-2,000	301.2	1.2778	1.2067	1.253	2.636	347.9
-1,500	297.9	1.207	1.1522	1.301	2.611	346
-1,000	294.7	1.1393	1.0996	1.352	2.585	344.1
-500	291.4	1.0748	1.0489	1.405	2.560	342.2
0	288.15	1.01325	1.0000	1.461	2.534	340.3
500	284.9	0.9546	0.9529	1.520	2.509	338.4
1,000	281.7	0.8988	0.9075	1.581	2.483	336.4
1,500	278.4	0.8456	0.8638	1.646	2.457	334.5
2,000	275.2	0.7950	0.8217	1.715	2.431	332.5
2,500	271.9	0.7469	0.7812	1.787	2.405	330.6
3,000	268.7	0.7012	0.7423	1.863	2.379	328.6
3,500	265.4	0.6578	0.7048	1.943	2.353	326.6
4,000	262.2	0.6166	0.6689	2.028	2.327	324.6
4,500	258.9	0.5775	0.6343	2.117	2.301	322.6
5,000	255.7	0.5405	0.6012	2.211	2.275	320.5
5,500	252.4	0.5054	0.5694	2.311	2.248	318.5
6,000	249.2	0.4722	0.5389	2.416	2.222	316.5
6,500	245.9	0.4408	0.5096	2.528	2.195	314.4
7,000	242.7	0.4111	0.4817	2.646	2.169	312.3
7,500	239.5	0.3830	0.4549	2.771	2.142	310.2
8,000	236.2	0.3565	0.4292	2.904	2.115	308.1
8,500	233.0	0.3315	0.4047	3.046	2.088	306.0
9,000	229.7	0.3080	0.3813	3.196	2.061	303.8
9,500	226.5	0.2858	0.3589	3.355	2.034	301.7
10,000	223.3	0.2650	0.3376	3.525	2.007	299.8

Elevation	Temperature	Pressure	Relative Density	Kinematic Viscosity	Thermal Conductivity	Speed of Sound
[m]	[K]	[bar]	[1]	[10^{-5} m^2 s^{-1}]	[10^{-2} Wm^{-1}K^{-1}]	[m/s]
h	T	P	ρ/ρ_o	ν	κ	a
10,500	220.0	0.2454	0.3172	3.706	1.980	297.4
11,000	216.8	0.2270	0.2978	3.899	1.953	295.2
11,500	216.7	0.2098	0.2755	4.213	1.952	295.1
12,000	216.7	0.1940	0.2546	4.557	1.952	295.1
12,500	216.7	0.1793	0.2354	4.930	1.952	295.1
13,000	216.7	0.1658	0.2176	5.333	1.952	295.1
13,500	216.7	0.1533	0.2012	5.768	1.952	295.1
14,000	216.7	0.1417	0.1860	6.239	1.952	295.1
14,500	216.7	0.1310	0.1720	6.749	1.952	295.1
15,000	216.7	0.1211	0.1590	7.300	1.952	295.1
15,500	216.7	0.1120	0.1470	7.895	1.952	295.1
16,000	216.7	0.1035	0.1359	8.540	1.952	295.1
16,500	216.7	0.09572	0.1256	9.237	1.952	295.1
17,000	216.7	0.08850	0.1162	9.990	1.952	295.1
17,500	216.7	0.08182	0.1074	10.805	1.952	295.1
18,000	216.7	0.07565	0.0993	11.686	1.952	295.1
18,500	216.7	0.06995	0.09182	12.639	1.952	295.1
19,000	216.7	0.06467	0.08489	13.670	1.952	295.1
19,500	216.7	0.05980	0.07850	14.784	1.952	295.1
20,000	216.7	0.05529	0.07258	15.989	1.952	295.1
22,000	218.6	0.04047	0.05266	22.201	1.968	296.4
24,000	220.6	0.02972	0.03832	30.743	1.985	297.7
26,000	222.5	0.02188	0.02797	42.439	2.001	299.1
28,000	224.5	0.01616	0.02047	58.405	2.018	300.4
30,000	226.5	0.01197	0.01503	80.134	2.034	301.7

Printed in Poland
by Amazon Fulfillment
Poland Sp. z o.o., Wrocław